The Violence of the
G R E E N
REVOLUTION

The Violence of the
GREEN
REVOLUTION

*Third World Agriculture, Ecology
and Politics*

VANDANA SHIVA

Zed Books Ltd.
London and New York

TWN
Third World Network
Penang, Malaysia

The Violence of the Green Revolution
is published by:
Zed Books Ltd., 7 Cynthia Street,
London N1 9JF, UK and Room 400,
175 Fifth Avenue, New York,
NY 10010,USA
and
Third World Network,
228 Macalister Road,
10400 Penang, Malaysia.

First Printing   : 1991
Second Printing : 1993
Third Printing : 1997

Cover design by Andrew Corbett

Printed by Jutaprint, 2 Solok Sungai Pinang 3,
11600 Penang, Malaysia.

ISBN: 0 86232 964 7 Hb
ISBN: 0 86232 965 5 Pb

Transferred to digital printing 2006

A catalogue record for this book is obtainable from the British Library.

Contents

Acknowledgements

This study of the ecological and social costs of the Green Revolution and the links between the ecological and ethnic crisis in Punjab, began in 1986, as part of the major project on "Conflicts Over Natural Resources" of the Peace and Global Transformation programme of the United Nations University.

Support from the Third World Network made it possible to continue the study during 1987-89. I am grateful to Prof. Uppal and Prof. Sidhu, and to the many scientists of Punjab Agricultural University in Ludhiana without whose inputs and cooperation, this study would not have been possible.

Anjali Kalgutkar has helped with maps and illustrations. The Third World Network secretariat especially Chee Yoke Heong, Christina Lim and Lim Jee Yuan, has taken care of all the details of the final editing and production of this book. I thank them for all of their contributions.

To Dr. Richaria,
for a lifetime of struggle to keep
people's agriculture knowledge alive.

Study area location within the Indian Union

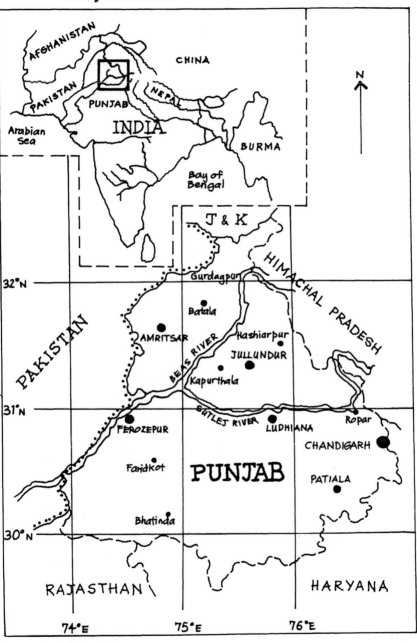

Source : Kang, 1982

INTRODUCTION

TWO MAJOR crises have emerged on an unprecedented scale in Asian societies during the 1980s. The first is the ecological crisis and the threat to life support systems posed by the destruction of natural resources like forests, land, water and genetic resources. The second is the cultural and ethnic crisis and the erosion of social structures that make cultural diversity and plurality possible as a democratic reality in a decentralised framework. The two crises are usually viewed as independent, both analytically as well as at the level of political action.

The tragedy of Punjab – of the thousands of innocent victims of violence over the past few years – has commonly been presented as an outcome of ethnic and communal conflict between two religious groups. This study presents a different aspect and interpretation of the Punjab tragedy. It introduces dimensions that have been neglected or gone unnoticed in understanding the emergent conflicts. It traces the conflicts and violence in contemporary Punjab to the ecological and political demands of the Green Revolution as an experiment in development and agricultural transformation. The Green Revolution has been heralded as a political and technological achievement, unprecedented in human history. It was designed as a techno-political strategy for peace, through the creation of abundance by breaking out of nature's limits and variabilities. Paradoxically, two decades of the Green Revolution have left Punjab ravaged by violence and ecological scarcity. Instead of abundance, Punjab has been left with diseased soils, pest-infested crops, water-logged deserts, and indebted and discontented farmers. Instead of peace, Punjab has inherited conflict and violence. 3,000 people were killed in Punjab during 1988. In 1987 the number was 1,544. In 1986, 598 people were killed.

The Punjab crisis is in large measure the tragic outcome of a resource intensive and politically and economically centralized experiment with food production. The experiment has failed – even though the Green Revolution miracle continues to be advertised on every platform of every agency that stood to gain from it, the Rockefeller and Ford Foundations, the World Bank, the seed and chemical multinationals, the Government of India and the various agencies it controls.

It is misleading to reduce the roots of communal conflict to religion, as most scholars and commentators have done, since the conflicts are also economic and political. They are not merely conflicts between two religious communities, but reflect cultural and social breakdown and tensions between a disillusioned farming community and a centralising state, which controls agricultural policy, finance, credit, inputs and prices of agricultural commodities. At the heart of these conflicts and disillusionments lies the Green Revolution. The present essay presents the other side of the Green Revolution story – its social and ecological costs hidden and hitherto unnoticed. In so doing, it also offers a different perspective on the multiple roots of ethnic and political violence. It illustrates that ecological and ethnic fragmentation and breakdown are intimately connected and are an intrinsic part of a policy of planned destruction of diversity in nature and culture to create the uniformity demanded by centralised management systems.

This book is also an attempt to understand the paradox that is today's Punjab. Statistics show Punjab to be India's most prosperous state, a model to which other regions and other countries aspire. Punjab's estimated gross domestic product per person was Rs2,528. The average per capita GDP for India was Rs1,334. The income of the average Punjabi was 65% greater than that of the 'average Indian'.

According to the 1981 census, Punjab's population was 16.7 million, a little less than 2.5% of India's population. Yet Punjab produces 7% of the country's foodgrains, has 10% of India's TV sets and 17% of India's tractors. It has three times more roads per square kilometer. An average Punjabi uses twice as much energy per hour than an average Indian and three times as such fertilizer per hectare. Compared to 28% of the national average 80% of Punjab's 50,400 square kilometers of land are irrigated. The average Punjabi has twice as much money in the bank as the average Indian. On all conventional indicators of progress and development, Punjab has done better than the rest of India.[1]

Yet Punjab is also the region most seething with discontent, with a sense of having been exploited and treated with discrimination. The sense of grievance runs so high that it has led to the largest numbers of killings in peacetime in independent India. At least 15,000 people have lost their lives in Punjab violence in the last six years.

The violence in contemporary Punjab goes against all conventional wisdom. The received view on societal violence identifies 'material scarcity' as the underlying determinant of man's inhumanity to man. From pre-neolithic times, it is argued, societal groups have always lived in environments too poor to satisfy their material needs.[2] Nature has thus been seen as a source of economic scarcity, scarcity has been seen as a source of conflict over scarce resources, and conflicts have in turn been seen as a source of violence.

'Development' then becomes a strategy to 'combat scarcity and dominate nature' to generate material abundance. This view of scarcity and of violence is shared by both the left and the right. Capital accumulation through appropriation of nature is seen by both ends of traditional political spectrum as a source of generating material abundance, and through it, conditions of peace. This orthodox view holds that 'the unprecedented control of the environment facilitated by a high-level technology, thus the possibility of eliminating toil and poverty is the necessary pre-requisite for overcoming the struggle between men themselves.'[3] The Green Revolution was conceived within this orthodox view of scarcity and violence. The Green Revolution was prescribed as a techno-politic strategy that would create abundance in agricultural societies and reduce the threat of communist insurgency and agrarian conflict. The British-American-sponsored Colombo Plan of 1952 was the explicit articulation of the development philosophy which saw the peasantry in Asia as incipient revolutionaries, who, if squeezed too hard, could be rallied against the politically and economically powerful groups. Rural development in general, and the Green Revolution in particular, assisted by foreign capital and planned by foreign experts, were prescribed as means for stabilising the rural areas politically 'which would include defusing the most explosive grievances of the more important elements in the countryside'.[4] This strategy was based on the idea of an agricultural revolution driven by scientific and technological innovations, since such an approach held the promise of changing the agrarian relations which had previously been politically so troublesome. Science and politics were wedded together in the very inception of the Green Revolution as a strategy for creating peace and prosperity in rural India.

However, after two decades, the invisible ecological, political and cultural costs of the Green Revolution have become visible. At the political level, the Green Revolution has turned out to be conflict-producing instead of conflict reducing. At the material level, production of high yields of commercial grain have generated new scarcities at the eco-system level, which in turn have generated new sources of conflict. It is in this multi-dimensional context of ecological and cultural disruption that an attempt will be made to understand the nature of Punjab violence – at the level of tacit and overt violence, at the level of real and perceived conflict, and at the level of ecological and political vulnerability and insecurity.

The ecological and ethnic crises in Punjab can be viewed as arising from a basic and unresolved conflict between the demands of diversity, decentralization and democracy on the one hand and the demands of uniformity, centralization, and militarization on the other. Control over nature and control over people were essential elements of the central-ised and centralising strategy of the Green Revolution. Eco-logical breakdown in nature and the political breakdown of society were essential implications of a policy based on tearing apart both nature and society. The Green Revolution was based on the assumption that technology is a superior substitute for nature, and hence a means of producing growth, unconstrained by nature's limits. Conceptually and empirically it is argued that the assumption of nature is a source of scarcity, and technology as a source of abundance, leads to the creation of technologies which create new scar-cities in nature through ecological destruction. The reduc-tion in availability of fertile land and genetic diversity of crops as a result of the Green Revolution practices indicates that at the ecological level, the Green Revolution produced scarcity, not abundance.

An attempt is also made to identify the many levels at which social and political insecurity was generated by the Green Revolution, and how, instead of stabilising and pacifying the countryside, it fueled a new pattern of conflict and violence. This includes an analysis of the communalisation of the Punjab conflicts which originally arose from the processes of political transformation associated with the Green Revolution. An attempt is made to anticipate the new levels of ecological, social and economic vulnerabilities that will arise from a second technological fix for Punjab in the form of the Pepsi project, which marks a new era of development policy in India.

Finally, the possibility and existence alternatives, in the context of deepening centralisation and control over agriculture, is discussed in the concluding chapter. Like Gandhi challenged the processes of colonisation linked with the first industrial revolution with the spinning wheel, peasants and Third World groups are challenging the recolonisation associated with the biotechnology revolution with their indigenous seeds.

The social and political planning that went into the Green Revolution aimed at engineering not just seeds but social relations as well. Punjab is an exemplar of how this engineering went out of control both at the material as well as the political level. Since this analysis is an attempt at grappling with the complex, and unanticipated factors unleashed by the Green Revolution, it avoids explanations based on deterministic and linear causality. Complex socio-ecological phenomena, even of intially conceived simplistically by paradigms of technological determistic framework of single cause-single effect. The best one can strive for is contextual causation, in which indications and suggestions are made of how the creation of certain contexts creates overwhelming conditions for certain processes to be unleashed. It is in this larger framework of invisible and unforeseen linkages that the roots of Punjab violence are traced to the ecological and political context of the Green Revolution.

References

1. Robin Jeffrey, *What is Happening to India*, London: Macmillan, 1986, p27.

2. Brian Easlea, *Science and sexual oppression*, London: Weidenfell and Nicholson, 1981, p8.

3. R Eccleshall, 'Technology and Liberation', *Radical philosophy*, No 11, Summer 1975, p9.

4. Robert S Anderson and Baker M Morrison, *Science, Politics and Agricultural Revolution in Asia*, Boulder: Westview Press, 1982, p7.

1

SCIENCE AND POLITICS
IN THE GREEN REVOLUTION

IN 1970, Norman Borlaug was awarded the Nobel Peace Prize for 'a new world situation with regard to nutrition...'. According to the Nobel Prize Committee, 'the kinds of grain which are the result of Dr Borlaug's work speed economic growth in general in the developing countries.'[1] The 'miracle seeds' that Borlaug had created were seen as a source of new abundance and peace. Science was awarded for having a magical ability to solve problems of material scarcity and violence.

'Green Revolution' is the name given to this science-based transformation of Third World agriculture, and the Indian Punjab was its most celebrated success. Paradoxically, after two decades of the Green Revolution, Punjab is neither a land of prosperity, nor peace. It is a region riddled with discontent and violence. Instead of abundance, Punjab has been left with diseased soils, pest-infested crops, waterlogged deserts and indebted and discontented farmers. Instead of peace, Punjab has inherited conflict and violence. At least 15,000 people have lost their lives in the last six years. 598 people were killed in violent conflict in

Punjab during 1986. In 1987 the number was 1544. In 1988, it had escalated to 3,000. And 1989 shows no sign of peace in Punjab.

The tragedy of Punjab – of the thousands of innocent victims of violence over the past five years – has commonly been presented as an outcome of ethnic and communal conflict between two religious groups. This study presents a different aspect and interpretation of the Punjab tragedy. It introduces dimensions that have been neglected or gone unnoticed in understanding the emergent conflicts. It traces aspects of the conflicts and violence in contemporary Punjab to the ecological and political demands of the Green Revolution as a scientific experiment in development and agricultural transformation. The Green Revolution has been heralded as a political and technological achievement, unprecedented in human history. It was designed as a strategy for peace, through the creation of abundance by breaking out of nature's limits and variabilities. In its very genesis, the science of the Green Revolution was put forward as a political project for creating a social order based on peace and stability. However, when violence was the outcome of social engineering, the domain of science was artificially insulated from the domain of politics and social processes. The science of the Green Revolution was offered as a 'miracle' recipe for prosperity. But when discontent and new scarcities emerged, science was delinked from economic processes.

On the one hand, contemporary society perceives itself as a science-based civilisation, with science providing both the logic as well as propulsion for social transformation. In this aspect science is self-consciously embedded in society.

On the other hand, unlike all other forms of social organisation and social production, science is placed **above** society. It cannot be judged, it cannot be questioned, it cannot be evaluated in the public domain.

As Harding has observed,

'Neither God nor tradition is privileged with the same credibility as scientific rationality in modern cultures... The project that science's sacredness makes taboo is the examination of science in just the ways any other institution or set of social practices can be examined.' [2]

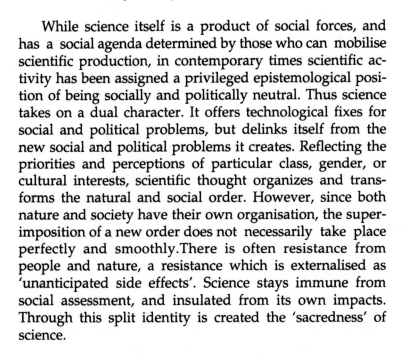

While science itself is a product of social forces, and has a social agenda determined by those who can mobilise scientific production, in contemporary times scientific activity has been assigned a privileged epistemological position of being socially and politically neutral. Thus science takes on a dual character. It offers technological fixes for social and political problems, but delinks itself from the new social and political problems it creates. Reflecting the priorities and perceptions of particular class, gender, or cultural interests, scientific thought organizes and transforms the natural and social order. However, since both nature and society have their own organisation, the superimposition of a new order does not necessarily take place perfectly and smoothly.There is often resistance from people and nature, a resistance which is externalised as 'unanticipated side effects'. Science stays immune from social assessment, and insulated from its own impacts. Through this split identity is created the 'sacredness' of science.

Within the structure of modern science itself are characteristics which prevent the perception of linkages. Fragmented into narrow disciplines and reductionist categories, scientific knowledge has a blind spot with respect to relational properties and relational impacts. It tends to decontextualise its own context. Through the process of decontextualisation, the negative and destructive impacts of science on nature and society are externalised and rendered invisible. Being separated from their material and political roots in the science system, new forms of scarcity and social conflict are then linked to other social systems e.g. religion.

The conventional model of science, technology and society locates sources of violence in politics and ethics, in the application of science and technology, not in scientific knowledge itself.[3] The assumed dichotomy between values and facts underlying this model implies a dichotomy between the world of values and the world of facts. In this view, sources of violence are located in the world of values while scientific knowledge inhabits the world of facts.

The fact-value dichotomy is a creation of modern reductionist science which,while being an epistemic response to a particular set of values, posits itself as independent of values. By splitting the world into fact vs values, it conceals the real difference between two kinds of value-laden facts. Modern reductionist science is characterised in the received view as the discovery of the properties and laws of nature in accordance with a 'scientific' method which generates claims of being 'objective', 'neutral' and 'universal'. This view of reductionist science as being a description of reality as it is, unprejudiced by value, is being rejected increasingly on historical and philosophical grounds. It has been historically established that all knowledge, including modern scientific knowledge, is built on the use of a plurality of methodologies, and reductionism itself is only one of the scientific options available.

The knowledge and power nexus is inherent to the reductionist system because the mechanistic order, as a conceptual framework, was associated with a set of values based on power which were compatible with the needs of commercial capitalism. It generates inequalities and domination by the way knowledge is generated and structured, the way it is legitimized, and by the way in which such knowledge transforms nature and society.

The experience of the Green Revolution in Punjab is an illustration of how contemporary scientific enterprise is politically and socially created, how it builds its immunity and blocks its social evaluation. It is an example of how science takes credit for successes and absolves itself from all responsibility for failures. The tragic story of Punjab is a tale of the exaggerated sense of modern science's power to control nature and society, and the total absence of a sense of responsibility for creating natural and social situations which are totally out of control. The externalization of the consequences of the Green Revolution from the scientific and technological package of the Green Revolution has been, in our view, a significant reason for the communalization of the Punjab crises.

It is, however, misleading to reduce the roots of the Punjab crisis to religion, as most scholars and commentators have done, since the conflicts are also rooted in the ecological, economic and political impacts of the Green Revolution.They are not merely conflicts between two religious communities, but reflect tensions between a disillusioned and discontented farming community and a centralising state, which controls agricultural policy, finance, credit, inputs and prices of agricultural commodities. At the heart of these conflicts and disillusionments lies the Green Revolution.

The present essay presents the other side of the Green Revolution story – its social and ecological costs hidden

and hitherto unnoticed. In so doing, it also offers a different perspective on the multiple roots of ethnic and political violence. It illustrates that ecological and ethnic fragmentation and breakdown are intimately connected and
are an intrinsic part of a policy of planned destruction of
diversity in nature and culture to create the uniformity
demanded by centralised management systems. The ecological and ethnic crises in Punjab can be viewed as arising
from a basic and unresolved conflict between the demands
of diversity, decentralisation and democracy on the one
hand, and the demands of uniformity, centralisation, and
militarisation on the other. Control over nature and control over people were essential elements of the centralised
and centralising strategy of the Green Revolution. Ecological breakdown in nature and the political breakdown of
society were consequences of a policy based on tearing
apart both nature and society.

The Green Revolution was based on the assumption
that technology is a superior substitute for nature, and
hence a means of producing limitless growth, unconstrained by nature's limits. However the assumption of
nature as a source of scarcity, and technology as a source
of abundance, leads to the creation of technologies which
create new scarcities in nature through ecological destruction. The reduction in availability of fertile land and genetic diversity of crops as a result of the Green Revolution
practices indicates that at the ecological level, the Green
Revolution produced scarcity, not abundance.

It was not just ecological insecurity but also social and
political insecurity was generated by the Green Revolution. Instead of stabilising and pacifying the countryside,
it fueled a new pattern of conflict and violence. The
communalisation of the Punjab conflicts which originally
arose from the processes of political transformation asso-

ciated with the Green Revolution, was based, in part, in externalising the political impacts of technological change from the domain of science and technology. A similar pattern of externalisation seems to be at play in the introduction of the 'biotechnology revolution', exemplified in Punjab by the Pepsi project.

The social and political planning that went into the Green Revolution aimed at engineering not just seeds but social relations as well. Punjab is an exemplar of how this engineering went out of control both at the material as well as the political level.

The Green Revolution and the Conquest of Nature

Half a century ago, Sir Alfred Howard, the father of modern sustainable farming wrote in his classic, *An Agricultural Testament*, that,

> '*In the agriculture of Asia we find ourselves confronted with a system of peasant farming which, in essentials, soon became stabilized. What is happening today in the small fields of India and China took place many centuries ago. The agricultural practices of the orient have passed the supreme test, they are almost as permanent as those of the primeval forest, of the prairie, or of the ocean.*' [4]

In 1889, Dr John Augustus Voelcker was deputed by the Secretary of State to India to advice the imperial government on the application of agricultural chemistry to Indian agriculture. In his report to the Royal Agricultural Society of England on the improvement of Indian Agriculture, Voelcker stated:

'I explain that I do not share the opinions which have been expressed as to Indian Agriculture being, as a whole, primitive and backward, but I believe that in many parts there is little or nothing that can be improved. Whilst where agriculture is manifestly inferior, it is more generally the result of the absence of facilities which exist in the better districts than from inherent bad systems of cultivation.... I may be bold to say that it is a much easier task to propose improvements in English agriculture than to make really valuable suggestions for that of India. To take the ordinary acts of husbandry, no where would one find better instances of keeping land scrupulously clean from weeds,of ingenuity in device of water raising appliances, of knowledge of soils and their capabilities as well as of the exact time to sow and to reap as one would in Indian agriculture, and this not at its best only but at its ordinary level. It is wonderful, too, how much is known of rotation, the system of mixed crops and of fallowing. Certain it is that I, at least, have never seen a more perfect picture of careful cultivation combined with hard labour, perseverance and fertility of resource.'[5]

When the best of western scientists were earlier sent to 'improve' Indian agriculture, they found nothing that could be improved in the principles of farming, which were based on preserving and building on nature's process and nature's patterns. Where Indian agriculture was less productive, it was not due to primitive principles or inferior practices, but due to interruptions in the flow of resources that made productivity possible. Land alienation, the reservation of forests and the expansion of cash crop cultivation were among the many factors, introduced during colonialism, that created a scarcity of local inputs of water and manure to maintain agricultural productivity.

In the second quarter of the century, from World War I to independence, Indian agriculture suffered a set-back as a consequence of complex factors including reduced exports due to worldwide recession, depression, and the near complete paralysis of shipping during World War II. The chaos of partition added to its decline, and the expansion of commercial crops like sugarcane and groundnuts, pushed food grains on to poorer lands where yields per acre were lower. The upheavals during this period left India faced with a severe food crisis.

There were two responses to the food crisis created through the war years and during partition. The first was indigenous, the second was exogenous. The indigenous response was rooted in the independence movement. It aimed at strengthening the ecological base of agriculture, and the self-reliance of the peasants of the country. *The Harijan*, a newspaper published by Mahatma Gandhi, which had been banned from 1942 to 1946, was full of articles written by Gandhi during 1946-1947 on how to deal with food scarcity politically, and by Mira Behn, Kumarappa and Pyarelal on how to grow more food using internal resources. On 10 June 1947, referring to the food problem at a prayer meeting Gandhi said:

'The first lesson we must learn is of self-help and self-reliance. If we assimilate this lesson,we shall at once free ourselves from disastrous dependence upon foreign countries and ultimate bankruptcy. This is not said in arrogance but as a matter of fact. We are not a small place, dependent for this food supply upon outside help. We are a sub-continent, a nation of nearly 400 millions. We are a country of mighty rivers and a rich variety of agricultural land, with inexhaustible cattle-wealth. That our cattle give much less milk than we need, is entirely our own fault. Our cattle-wealth is any day capable of

*giving us all the milk we need. Our country, if it had
not been neglected during the past few centuries, should
not today only be providing herself with sufficient food,
but also be playing a useful role in supplying the outside
world with much-needed foodstuffs of which the late war
has unfortunately left practically the whole world in
want. This does not exclude India.'*[6]

Recognising that the crisis in agriculture was related
to a breakdown of nature's processes, India's first agriculture minister, K M Munshi, had worked out a detailed
strategy on rebuilding and regenerating the ecological base
of productivity in agriculture based on a bottom-up
decentralised and participatory methodology.

In a seminar on 27 September 1951, organised by the
Agriculture Ministry, a program of regeneration of Indian
Agriculture was worked out, with the recognition that the
diversity of India's soils, crops and climates, had to be
taken into account. The need to plan from the bottom, to
consider every individual village and sometimes every
individual field was considered essential for the programme
called 'land transformation'. At this seminar, K M Munshi
told the State Directors of Agricultural extension:

*'Study the Life's Cycle in the village under your charge
in both its aspects – hydrological and nutritional. Find
out where the cycle has been disturbed and estimate the
steps necessary for restoring it. Work out the village in
four of its aspects, (1) existing conditions, (2) steps necessary for completing the hydrological cycle, (3) steps
necessary to complete the nutritional cycle, and a complete picture of the village when the cycle is restored,
and (4) have faith in yourself and the programme. Nothing is too mean and nothing too difficult for the man
who believes that the restoration of the life's cycle is not*

only essential for freedom and happiness of India but is essential for her very existence.[7]

Repairing nature's cycles and working in partnership with nature's processes was viewed as central to the indigenous agricultural policy.

However, while Indian scientists and policy-makers were working out self-reliant and ecological alternatives for the regeneration of agriculture in India, another vision of agricultural development was taking shape in American foundations and aid agencies. This vision was based not on cooperation with nature, but on its conquest. It was based not on the intensification of nature's processes, but on the intensification of credit and purchased inputs like chemical fertilizers and pesticides. It was based not on self-reliance, but dependence. It was based not on diversity but uniformity. Advisors and experts came from America to shift India's agricultural research and agricultural policy from an indigenous and ecological model to an exogenous, and high input one, finding, of course, partners in sections of the elite, because the new model suited their political priorities and interests.

There were three groups of international agencies involved in transferring the American model of agriculture to India – the private American Foundations, the American Government and the World Bank. The Ford Foundation had been involved in training and agricultural extension since 1952. The Rockefeller Foundation had been involved in remodelling the agricultural research system in India since 1953. In 1958, the Indian Agricultural Research Institute which had been set up in 1905 was reorganised, and Ralph Cummings, the field director of the Rockefeller Foundation, became its first dean. In 1960, he was succeeded by A B Joshi, and in 1965 by M S Swaminathan.

Besides reorganising Indian research institutes on American lines, the Rockefeller Foundation also financed the trips of Indians to American institutions. Between 1956 and 1970, 90 short-term travel grants were awarded to Indian leaders to see the American agricultural institutes and experimental stations. One hundred and fifteen trainees finished studies under the Foundation. Another 2,000 Indians were financed by USAID to visit the US for agricultural education during the period. The work of the Rockefeller and Ford Foundations was facilitated by agencies like the World Bank which provided the credit to introduce a capital intensive agricultural model in a poor country. In the mid-1960s India was forced to devalue its currency to the extent of 37.5%. The World Bank and USAID also exerted pressure for favourable conditions for foreign investment in India's fertilizer industry, import liberalisation, and elimination of domestic controls. The World Bank provided credit for the foreign exchange needed to implement these policies. The foreign exchange component of the Green Revolution strategy, over the five year plan period (1966-71) was projected to be Rs1,114 crores, which converted to about $2.8 billion at the then official rate. This was a little over six times the total amount allocated to agriculture during the preceding third plan (Rs 191 crores). Most of the foreign exchange was needed for the import of fertilizers, seeds and pesticides, the new inputs in a chemically intensive strategy. The World Bank and USAID stepped in to provide the financial input for a technology package that the Ford and Rockefeller Foundations had evolved and transferred.

Within India, the main supporter of the Green Revolution strategy was C Subramaniam, who became agriculture minister in 1964, and M S Swaminathan, who became the Director of IARI in 1965, and had been trained by Norman Borlaug, who worked for Rockefeller's agricultural programme in Mexico. After a trip to India in 1963, he despatched 400kg of semidwarf varieties to be tested in India. In 1964, rice seeds were brought in from the International Rice Research Institute (IRRI) in the Philippines (which had recently been set up with Ford and Rockefeller funds). In the same year Ralph Cummings felt that sufficient testing had been done to release the varieties on a large scale. He approached C Subramaniam to see if the new agriculture minister would be willing to throw his support to accelerating the process of introducing the green revolution seeds. Subramaniam acknowledges that he decided to follow Cumming's advice quickly, and began to formulate a strategy for using the new varieties.[8]

Others in India were not as willing to adopt the American agricultural strategy. The Planning Commission was concerned about the foreign exchange costs of importing the fertilizer needed for application to the HYV's in a period of a severe balance-of-payments crisis. Leading economists B S Minhas and T S Srinivas questioned the strategy on economic grounds. State governments worried that adoption of the new seeds would reduce their autonomy in agricultural research. Agricultural scientist objected to the new varieties for risks to disease and displacement of small peasants. The only group supporting Subramaniam were the younger agricultural scientists trained over the past decade in the American paradigm of agriculture.

The occurrence of drought in 1966 caused a severe drop in food production in India, and an unprecedented increase in food grain supply from the US. Food dependency was used to set new policy conditions on India. The US President, Lyndon Johnson, put wheat supplies on a short tether. He refused to commit food aid beyond one month in advance until an agreement to adopt the Green

Revolution package was signed between the Indian agriculture minister, C S Subramaniam and the US Secretary of agriculture, Orville Freeman.[9]

Lal Bahadur Shastri, the Indian Prime Minister in 1965 had raised caution against the rushing into a new agriculture based on new varieties. With his sudden death in 1966 the new strategy was more easy to introduce. The Planning Commission, which approves all large investment in India, was also bypassed since it was viewed as a bottleneck.

Rockefeller agricultural scientists saw Third World farmers and scientists as not having the ability to improve their own agriculture. They believed that the answer to greater productivity lay in the American-styled agricultural system. However the imposition of the American model of agriculture did not go unchallenged in the Third World or in America. Edmundo Taboada, who was head of the Mexican office of Experiment Stations, maintained, like K M Munshi in India, that ecologically and socially appropriate research strategies could only evolve with the active participation of the peasantry.

> 'Scientific Research must take into account the men that will apply its results... Perhaps a discovery may be made in the laboratory, a greenhouse or an experimental station, but useful science, a science that can be applied and handled must emerge from the local laboratories of..... small farmers, ejidatorios and local communities.'[10]

Together peasants and scientists searched for ways to improve the quality of 'criollo' seeds (open pollinated indigenous varieties) which could be reproduced in peasant fields. However, by 1945, the Special Studies Bureau in the Mexican Agriculture Ministry, funded and administered by the Rockefeller Foundation, had eclipsed the indigenous research strategy and started to export to Mexico the American agricultural revolution. In 1961, the Rockefeller-financed center took the name of CIMMYT (Centro International de Mejoramiento de Maiz Y Trigo or the International Maize and Wheat Improvement Centre). The American strategy, reinvented in Mexico, then came to the entire Third World as the 'Green Revolution'.

The American model of agriculture had not done too well in America, though its nonsustainability and high ecological costs went ignored. The intensive use of artificial fertilizers, extensive practice of monocultures, and intensive and extensive mechanisation had turned fertile tracts of the American prairies into a desert in less than thirty years.

The American Dust Bowl of the 1930s was in large measure a creation of the American agricultural revolution. Hyman reports,

'When, between 1889 and 1900, thousands of farmers were settling in Oklahoma, it must have seemed to them that they were founding a new agricultural civilization which might endure as long as Egypt. The grandsons, and even the sons of these settlers who so swiftly became a disease of their soil, trekked from their ruined farmsteads, their buried or uprooted crops, their dead soil, with the dust of their own making in their eyes and hair, the barren sand of a once fertile plain gritting between their teeth.. The pitiful procession passed westward, an object of disgust – the God-dam'd Okies. But these God-dam'd okies were the scapegoats of a generation, and the God who had damned them was perhaps after all a Goddess, her name Ceres, Demeter, Maia, or something older and more terrible. And what she damned

them for was their corruption, their fundamental igno-
rance of the nature of her world, their defiance of the
laws of co-operation and return which are the basis of
life on this planet'.[11]

When an attempt was made to spread this ecologically devastating vision of agriculture to other parts of the world through Rockefeller Foundation programmes, notes of caution were sounded.

The American strategy of the Rockefeller and Ford Foundations differed from the indigenous strategies primarily in the lack of respect for nature's processes and people's knowledge. In mistakenly identifying the sustainable and lasting as backward and primitive, and in perceiving nature's limits as constraints on productivity that had to be removed, American experts spread ecologically destructive and unsustainable agricultural practices worldwide. The Ford Foundation had been involved in agricultural development in India since 1951. In 1952, 15 community development projects, each covering about 100 villages, were started, with Ford Foundation financial assistance. This programme was however shed in 1959 when a Ford Foundation mission of thirteen North American agronomists to India argued that it was impossible to make simultaneous headway in all of India's 550,000 villages. Their recommendations for a selective and intensive approach among farmers and among districts led to the winding down of the community development programme and the launching of the Intensive Agricultural Development Programme (IADP) in 1960-61.

The IADP totally replaced an indigenous, bottom up, organic-based strategy for regenerating Indian agriculture, with an exogenous, top down chemically intensive one. Industrial inputs like chemical fertilizers and pesticides were seen as breaking Indian agriculture out of the 'shackles of the past', as an article on 'Ford Foundations Involvement in Intensive Agricultural Development in India' stated:

> 'India is richly endowed with sunshine, vast land areas (much of it with soils responsive to modernizing farming), a long growing season (365 days a year in most areas). Yet the solar energy, soil resources, crop growing days and water for irrigation are seriously underused or misused. India's soils and climate are among the most underused in the world. Can multiple cropping help Indian farmers utilize these vast resources more effectively – the answer must be yes.'

> 'New opportunities for intensifying agricultural programs through multiple cropping are presenting themselves; led by the plant breeder there are new short season, fertilizer responsive, non-photo sensitive crops and varieties that under skillful farming practices have high yield potential; chemical fertilizer supplies are increasing rapidly – this frees the Indian cultivator from the shackles of the past permitting only very modest improvement of soil fertility through green manure and compost and the slow, natural recharge of soil nutrients. Also, up until recently varieties were bred for these conditions, plant protection was applied after the damage was done, and so on – a status quo agriculture. This has changed. Indian farmers are prepared to innovate and change; Indian leaders in agricultural development, extension, research and administration are beginning to understand the new potentials; intensive agriculture, first identified under IADP, is now India's food production strategy.' [12]

Under the Ford Foundation programme, agriculture

was transformed from one that is based on internal inputs that are easily available at no costs, to one that is dependent on external inputs for which credits became necessary. Instead of promoting the importance of agriculture in all regions, the IADP showed favouritism to specially selected areas for agricultural development, to which material and financial resources of the entire country were diverted. The latter however was a failed strategy where native varieties of food crops were concerned. The native crops tend to 'lodge' or fall under the intensive application of chemical fertilizers, thus putting a limit to fertilize use.

As a spokesman of the Ford Foundation put it, 'The programme revealed the urgent need for improved crop varieties as it was found that the native varieties (the only ones available during these early years) responded very poorly to improved practices and produced low yields even when subjected to other modern recommended practices.'

It was not that native crop varieties were low yielding inherently. The problem with indigenous seeds was that they could not be used to consume high doses of chemicals. The Green Revolution seeds were designed to overcome the limits placed on chemically intensive agriculture by the indigenous seeds. The new seeds thus became central to breaking out of nature's limits and cycles. The 'miracle' seeds were therefore at the heart of the science of the 'Green Revolution'.

The combination of science and politics in creating the Green Revolution goes back to the period in the 1940s when Daniels, the US Ambassador to the Government of Mexico, and Henry Wallace, Vice-President of the United States set up a scientific mission to assist in the development of agricultural technology in Mexico. The office of Special Studies was set-up in Mexico in 1943 within the agricultural ministry as a co-operation venture between the Rockefeller Foundation and the Mexican Government. In 1944, Dr J George Harrar, head of the new Mexican research programme and Dr Frank Hanson, an official of the Rockefeller Foundation in New York invited Norman Borlaug to shift from his classified wartime laboratory job in Dupont to the plant breeding programme in Mexico. By 1954, Borlaug's 'miracle seeds' of dwarf varieties of wheat had been bred. In 1970, Borlaug had been awarded the 'Nobel Peace Prize' for his 'great contribution towards creating a new world situation with regard to nutrition... The kinds of grain which are the result of Dr Borlaug's work speed economic growth in general in the developing countries.'[13]

This assumed link between the new seeds and abundance, and between abundance and peace was sought to be replicated rapidly in other regions of the world, especially Asia.

Impressed with the successful diffusion of 'miracles' seeds of wheat from CIMMYT (International Maize & Wheat Improvement Centre) which had been set-up in 1956 on the basis of the Rockefeller Foundation and Mexican Government programme, the Rockefeller and Ford Foundations in 1960 established IRRI, the International Rice Research Institute in the Philippines, which by 1966 was producing 'miracle' rice, to join the 'miracle' wheats from CIMMYT.

CIMMYT and IRRI were the international agricultural research centres which grew out of the Rockefeller Foundation country programme to launch the new seeds and the new agriculture across Latin America and Asia.By 1969, the Rockefeller Foundation in co-operation with the Ford Foundation had established the Centro International de

Table 1.1 : The Consultative Group On International Agricultural Research (CGIAR) System 1984

Acronym (year established)	Center	Location	1984 budget (millions of dollars)
IRRI(1960)	International Rice Research Institute	Los Banos,	22.5
CIMMYT(1966)	Centro Internacional de Mejoramientio Maizy Trigo	Mexico City, Mexico	21.0
IITA(1967)	International Institute of Tropical Agriculture	Ibadan, Nigeria	21.2
CIAT(1968)	Centro Internacional de Agricultura Tropical	Cali, Colombia	23.1
CIP(1971)	Centro Internacional de la papa	Lima, Peru	10.9
WARDA(1971)	West African Rice Development Association	Monrovia, Liberia	2.9
ICRISAT(1972)	International Crops Research Institute for the Semi-Arid Tropics	Hyderabad, India	22.1
ILRAD(1973)	International Laboratory for Research for Animal Diseases	Nairobi, Kenya	9.7
IBPGR(1974)	International board for Plant Genetic Resources	Rome, Italy	3.7
ILCA(1974)	International Livestock Center for Africa	Addis Ababa, Ethiopia	12.7
IFPRI(1975)	International Food Policy Research Institute	Washington DC, USA	4.2
ICARDA(1976)	International Center for Agricultural Research in the Dry Areas	Aleppo, Syria	20.4
ISNAR(1980)	International Service for National Agricultural Research	The Hague Netherlands	3.5

Source: Consultative Group on International Agricultural Research, Washington, DC 1984

Table 1.2 : Membership of the Consultative Group on International Agricultural Research (January 1983)

Countries	International organizations	Foundations	Fixed-term members representing developing countries
Australia	African Development Bank	Ford Foundation	Asian region: Indonesia and
Belgium	Arab Fund for Economic	Research and development	Pakistan
Brazil	Development	Research Centre	African region: Senegal and Tanzania
Canada	Asian Development Bank	Kellogg Foundation	
Denmark	Commission of the European	Leverhulme Trust	Latin American region:Colombia
France	Communities	Rockefeller Foundation	and Cuba
Germany	Food and Agriculture organization of		Southern and Eastern European
India	the United Nations		region: Greece and Romania
Ireland	Inter-American Development Bank		Near Eastern region: Iraq and
Italy	International Bank for Reconstruction and Development		Libya
Japan			
Mexico	OPEC Fund		
Netherlands	United Nations Development Programme		
Nigeria			
Norway	United Nations Environment Programme		
Philippines			
Saudi Arabia			
Spain			
Sweden			
Switzerland			
United Kingdom			
United States			

Source: *Consultative Group on International Agricultural Research, 1818H. Street NW, Washington DC 20433, USA.*

Table 1.3 : IRRI finances according to source
(1961-1980) (US dollars)

Contributor	Amount	% of total	Year(s) of Grant
Ford Foundation	23,950,469	18.84	1961-80
Rockefeller Foundation	20.460,431	16.1	1961-80
US AID	28,982,114	22.80	1967-80
International Organizations	20,334,788	16	
Asian Development Bank	800,000		1975,1977
European Economic Community	3,011,219		1978-80
Fertilizer Development Center	70,939		1979-80
Foundation for International Potash Research	7,375		1963-65
International Board for plant Genetic Resources	208.100		1977,1979-80
International Center of Insect Physiology and Ecology	125,432		1978-80
International Development Research Center	3,710,736		1972-73, 1975-76, 1978-80
International Development Association	7,775,000		1973-80
International Fund for Agricultural Research	500,000		1980
International Potash Institute/ Potash Institute of North America	68,064		1963,1965-66 1968-69, 1971-79
Fertilizer Development Center	70,939		1979-80
OPEC Special Fund	200,000		1980
UN Economic and Social Commission	6,000		1970,1979
UN Food and Agriculture organization (FAO)	2,650		1969
UN Environment program	280,000		1974-78
UN Development Program	3,559,273		1974-78,1978
World Phosphate Rock Institute	10,000		1975
National Governments	31,920,619	25.11	
Australia	4,185,459		1975-80
Belgium	148,677		1977

Canada	6,507,862		1974-80
Denmark	443,048		1978-80
Federal Republic of Germany	3,459,159		1974-80
Indonesia	1,619,119		1973-80
Iran	250,000		1977
Japan	8,882,145		1971-77, 1979-80
Korea	82,259		1980
The Netherlands	1,168,673		1971-79
New Zealand	137,450		1973,1976-78
Philippines	100,000		1980
Saudi Arabia	274,300		1976-77,1980
Sweden	285,700		1979-80
United Kingdom	4,073,824		1973-76, 1979-80
Corporations	345,726	0.27	
Bayer	9,333		1971,1973
Boots Company	1,000		1977
Chevron Chemicals	2,993		1972,1977
Ciba-Geigy	20,500		1968,1970, 1972, 1975 1978-80
Cyanamid	19,000		1975-76,1978, 1980
Dow Chemical	10,153		1967-70
Eli Lilly & Co (ELANCO)	6,000		1968-70
Esso Engineering and Research Company	4,406		1964-68
FMC	9,000		1975-77,1980
Gulf Research and Development Company	3,500		1969,1972
Hoechst	11,891		1972,1975-78, 1978
Imperial Chemical Industries	55,000		1967,69, 1971-76 1979,80
International Business Machines Corp. (IBM)	7,000		1967
International Minerals and Chemical Corp.	60,000		1966-67,1975
Kemanober	500		1980
Minnesota Mining and Manufacturing Company	1,000		1974

Manufacturing Company			
Monsanto	12,500		1967,1969, 1971-72, 1976,1978-80
Montedison	8,982		1977-78,1980
Occidental Chemical	500		1971
Pittsburg plate Glass Co.	2,000		1967
Plant Protection Ltd	5,000		1966
Shell Chemical Company	42,672		1969-70, 1972-73 1975,1977-78 1980
Stauffer Chemical Company	40,000		1967-69, 1971-76 1978-80
Union Carbide	11,000		1968,1970
Uniroyal Chemical	496		1980
Upjohn	1,200		1972
Government agencies	1,030,872	0.81	
National Institute of Health(US)	383,708		1978-80
National Food and Agriculture Council (Philippines)	276,859		1973,1976-80
National Science Development Board (Philippines)	104,172		1963,1965, 1967-68,1973, 1975-76,1964- 68,1976
Philippine Council for Agriculture Resources and Research	198,911		1976-80
Universities	13,634	0.01	
East-West Center (Hawaii)	1,500		1976,1978
University Hohenheim (Stuttgart)	4,370		1980
United Nations University	7,764		1980
Others	61,557	0.05	1966,1969, 1977
Total	127,100,210		

Source: *International Rice Research Institute, Annual Report from 1962-1980.*

Agriculture Tropical (CIAT) in Columbia and the International Institute for Tropical Agriculture (IITA) in Nigeria.

In 1971, at the initiative of Robert McNamara, the President of the World Bank, a Consultative Group on International Agricultural Research (CGIAR) was formed to finance the network of these international agricultural centres (IARC). Since 1971, nine more IARC's were added to the CGIAR system. The International Crops Research Institute for the Semi-Arid Tropics (ICRISAT) was started in Hyderabad in India in 1971. The International Laboratory for Research on Animal Diseases (ICRAD) and the International Livestock Centre for Africa (ILCA) were approved in 1973.

The Consultative Group had 16 donors, who contributed $20.06 million in 1972. By 1981, the budget had shot up to $157.945 million provided by 40 donors.

The growth of the international institutes was based on the erosion of the decentralised knowledge systems of Third World peasants and Third World research institutes. The centralised control of knowledge and genetic resources was, as mentioned, not achieved without resistance. In Mexico, peasant unions protested against it. Students and professors at Mexico's National Agricultural College in Chapingo went on strike to demand a programme different from the one that emerged from the American strategy and was more suitable to the small-scale poor farmers and to the diversity of Mexican agriculture.

The International Rice Research Institute was set up in 1960 by the Rockefeller and Ford Foundations, nine years after the establishment of a premier Indian Institute, the Central Rice Research Institute (CRRI) in Cuttack. The Cuttack institute was working on rice research based on indigenous knowledge and genetic resources, a strategy clearly in conflict with the American-controlled strategy of the International Rice Research Institute. Under international pressure, the director of CRRI was removed when he resisted handing over his collection of rice germplasm to IRRI, and when he asked for restraint in the hurried introduction of the HYV (High Yield Varieties) from IRRI.

The Madhya Pradesh government gave a small stipend to the ex-director of CRRI so that he could continue his work at the Madhya Pradesh Rice Research Institute (MPRRI) at Raipur. On this shoestring budget, he conserved 20,000 indigenous rice varieties in situ in India's rice bowl in Chattisgarh. Later the MPRRI, which was doing pioneering work in developing a high yielding strategy based on the indigenous knowledge of the Chattisgarh tribals, was also closed down due to pressure from the World Bank (which was linked to IRRI through CGIAR) because MPRRI had reservations about sending its collection of germplasm to IRRI.[14]

In the Philippines, IRRI seeds were called 'Seeds of Imperialism'. Robert Onate, president of the Philippines Agricultural Economics and Development Association observed that IRRI practices had created debt and a new dependence on agrichemicals and seeds. 'This is the Green Revolution connection', he remarked. 'New seeds from the CGIAR global crop/seed systems which will depend on the fertilizers, agrichemicals and machineries produced by conglomerates of transnational corporations.'[15]

Centralism of knowledge was built into the chain of CGIAR's from which technology was transferred to second-order national research centres. The diverse knowledge of local cultivators and plant breeders was displaced. Uniformity and vulnerability were built into international research centres run by American and American-trained experts breeding a small set of new varieties that would displace the thousands of locally cultivated plants in the agricultural systems, built up over generations on the basis on knowledge generated over centuries.

Politics was built into the Green Revolution because the technologies created were directed at capital intensive inputs for best endowed farmers in the best endowed areas, and directed away from resource prudent options of the small farmer in resource scarce regions. The science and technology of the Green Revolution excluded poor regions and poor people as well as sustainable options. American advisors gave the slogan of 'building on the best'. The science of the Green Revolution was thus essentially a political choice.

As Lappe and Collins have stated:

'Historically, the Green Revolution represented a choice to breed seed varieties that produce high yields under optimum conditions. It was a choice ***not*** *to start by developing seeds better able to with stand drought or pests. It was a choice* ***not*** *to concentrate first on improving traditional methods of increasing yields, such as mixed cropping. It was a choice* ***not*** *to develop technology that was productive, labour-intensive, and independent of foreign input supply. It was a choice* ***not*** *to concentrate on reinforcing the balanced, traditional diets of grain plus legumes.'*[16]

The crop and varietal diversity of indigenous agriculture was replaced by a narrow genetic base and monocultures. The focus was on internationally traded grains, and a strategy of eliminating mixed and rotational cropping, and diverse varieties by varietal simplicity. While the new varieties reduced diversity, they increased resource use of water, and of chemical inputs such as pesticides and fertilizers.

The strategy of the Green Revolution was aimed at transcending scarcity and creating abundance. Yet it put new demands on scarce renewable resources and generated new demands for non-renewable resources. The Green Revolution technology require heavy investments in fertilizers, pesticides, seed, water and energy. Intensive agriculture generated severe ecological destruction, and created new kinds of scarcity and vulnerability, and new levels of inefficiency in resource use. Instead of transcending the limits put by natural endowments of land and water, the Green Revolution introduced new constraints on agriculture by wasting and destroying land, water resources, and crop diversity. The Green Revolution had been offered as a miracle.

Yet, as Angus Wright has observed:

> 'One way in which agricultural research went wrong was precisely in saying and allowing it to be said that some miracle was being produced.... Historically, science and technology made their first advances by rejecting the idea of miracles in the natural world. Perhaps it would be best to return to that position.'[17]

The Green Revolution and the Control of Society

The Green Revolution was promoted as a strategy that would simultaneously create material abundance in agricultural societies and reduce agrarian conflict. The new seeds of the Green Revolution were to be seeds of plenty and were also to be the seeds of a new political economy in Asia.

The Green Revolution was necessarily paradoxical. On the one hand it offered technology as a substitute to both nature and politics, in the creation of abundance and peace. On the other hand, the technology itself demanded more intensive natural resource use along with intensive external inputs and involved a restructuring of the way power was distributed in society. While treating nature and politics as dispensable elements in agricultural transformation, the Green Revolution created major changes in natural ecosystems and agrarian structures. New relationships between science and agriculture defined new links between the state and cultivators, between international interests and local communities, and within the agrarian society.

The Green Revolution was not the only strategy available. There was another strategy for agrarian peace based on reestablishing justice through land reform and the removal of political polarisation which was at the base of political unrest in agrarian societies.

Colonialism had dispossessed peasants throughout the Third World of their entitlements to land and to a full participation in agricultural production. In India, the British introduced the system of 'Zamindari' or landlordship, to help divert land from growing food to growing opium and indigo, as well as to extract revenue from the cultivators.

R P Dutt records the sudden increase in agricultural revenues when the East India Company of British soldier-traders took over revenue rights of Bengal:

'In the last year of administration of the last Indian

*ruler of Bengal, in 1764-65, the land revenue realized
was £817,000. In the first year of the company's administration, in 1765-66, the land revenue realized in
Bengal was £1,470,000. By 1771-2, it was £2,348,000
and by 1775-6, it was £2,818,000. When Lord Cornwal-
lis fixed the permanent settlement in 1793, he fixed
£3,400,000.'* [18]*

The diversion of increasing amounts of agricultural
produce as a source of colonial revenue took its toll in
terms of deteriorating conditions of peasants and agricultural production.

According to Bajaj:

*'With more and more money flowing into the British
hands the village and the producer were left with precious little to feed themselves and maintain the various
village institutions that catered to their needs. According to Dharampal's estimates, whereas around 1750, for
every 1000 units of produce the producer paid 300 as
revenue, only 50 of which went out to the central authority, the rest remaining within the village; by 1830,
he had to give away 650 units as revenue, 590 of which
went straight to the central authority. As a result of this
level of revenue collection the cultivators and the villag-
ers both were destroyed.'* [19]*

In Mexico, the Spanish instituted the 'hacienda' (large
estate) owners. After two centuries of colonisation, haciendas dominated the countryside. They covered 70 million
hectares of the land, leaving only 18 million hectares under
the control of indigenous communities. According to
Esteva, by 1910, around 8,000 'haciendas' were in the hands
of a small number of owners, occupying 113 million hectares, with 4,500 managers, 300,000 tenants, and 3,000,000

indentured peons and sharecroppers. An estimated 150,000 'Indian' communal land holders occupied 6 million hectares. Less than 1% of the population owned over 90% of the land and over 90% of the rural population lacked any access to it. [20]

Between 1910 and 1917, over one million peasants in Mexico had died fighting for land. Between 1934 and 1949, Lazaro Cardenas redistributed 78 million acres and benefited 42% of the entire agricultural population. Under the new distribution small farmers owned 47% of the land.

As Lappe and Collins report:

'Social and economic process was being achieved not through dependence on foreign expertise or costly imported agricultural inputs but rather with the abundant, underutilized resources of local peasants. While production increases were seen as important, the goal was to achieve them through helping every peasant to be productive, for only then would the rural majority benefit from the production increases. Freed from the fear of landlords, bosses, and money-lenders, peasants were motivated to produce, knowing that at last they would benefit from their own labor. Power was perceptibly shifting to agrarian reform organizations controlled by those who worked the fields.' [21]

The result of this gain in political and economic power of the peasants was the erosion of power of the powerful hacienda owners, and the US corporate sector whose investment dropped by about 40% between the mid thirties and early 1940s.

When Cardenas was succeeded by Avila Camacho, a fundamental shift was induced in Mexico's agricultural policy. It was now to be guided by American control over research and resources for agriculture through the Green Revolution Strategy.

Peasant movements had tried to restructure agrarian relationships through the recovery of land rights. The Green Revolution tried to restructure social relationships by separating issues of agricultural production from issues of justice. Green Revolution politics was primarily a politics of depoliticisation.

According to Anderson and Morrison:

'The founding of the International Rice Research Institute in Los Banos in 1960 was the institutional embodiment of the conviction that high quality agricultural research and its technological extensions would increase rice production, ease the food supply situation, spread commercial prosperity in the rural areas, and defuse agrarian radicalism.'[22]

In the 1950s, the newly independent countries of Asia were faced with rising peasant unrest. When the Chinese Communist Party came to power it had encouraged local peasants' associations to seize land, cancel debts and redistribute wealth. Peasant movements, inspired by the Chinese experience, flared up in the Philippines, Indonesia, Malaysia, Vietnam and India. The new political authorities in these Asian countries had to find a means to control agrarian unrest and stabilize the political situation. This 'would include defusing the most explosive grievances of the more important elements in the countryside.'[23]

In India land reforms had been viewed as a political necessity at the time of independence. Most states had initiated land reforms by 1950 in the form of abolition of

Zamindari (landlordism), security of tenure for tenant cultivators and fixation of reasonable rents. Ceilings on land holdings were also introduced. In spite of weaknesses in the application of land-reform strategies, they provided relief to the cultivators through the 1950s and 1960s. Aggregate crop output kept increasing during the 1950s in response to the restoration of some just order in land-relations.

A second strategy for agricultural production and agrarian peace was however being worked out internationally, driven by concern at the 'loss of China'. American agencies like the Rockefeller and Ford Foundations, US Aid, the World Bank etc. mobilised themselves for a new era of political intervention.

As Anderson and Morrison have observed:

'Running through all these measures, whether major or minor in their effect, was the concern to stabilize the countryside politically. It was recognised internationally that the peasantry were incipient revolutionaries and if squeezed too hard could be rallied against the new bourgeois-dominated governments in Asia.This recognition led many of the new Asian governments to join the British-American-sponsored Colombo Plan in 1952 which explicitly set out to improve conditions in rural Asia as a means of defusing the Communist appeal. Rural development assisted by foreign capital was prescribed as a means of stabilizing the countryside.'[24]

In Cleaver's view:

'Food was clearly recognised as a political weapon in the efforts to thwart peasant revolution in many places in Asia... from its beginning the development of the Green

Revolution grains constituted mobilizing science and technology in the service of counter-revolution.'[25]

Science and politics were thus wedded together in the very inception of the Green Revolution as a strategy for increasing material prosperity and hence defusing agrarian unrest. For the social planners in national governments and international aid agencies, the Science and Technology of the Green Revolution were an integral part of sociopolitical strategy aimed at pacifying the rural areas of developing nations in Asia, not through redistributive justice but through economic growth. And agriculture was to be the source of this new growth.

While the Green Revolution was clearly political in reorganising agricultural systems, the concern for political issues such as participation and equity, was consciously by-passed and was replaced by the political concern for stability. Goals of growth had to be separated from goals of political participation.

As David Hopper, then with Rockefeller Foundation wrote in his 'Strategy for the Conquest of Hunger':

'Let me begin my examination of the essentials for pay-off by focussing on public policy for agricultural growth. The confusion of goals that has characterized purposive activity for agricultural development in the past cannot persist if hunger is to be overcome. National governments must clearly separate the goal of growth from the goals of social development and political participation....These goals are not necessarily incompatible, but their joint pursuit in unitary action programs is incompatible with development of an effective strategy for abundance. To conquer hunger is a large task. To ensure social equity and opportunity is another

Table 1.4: Compound Rates of Growth

	Production		Area		Yield (percent per annum)	
Period	(a) 1949-50 to 1964-65	(b) 1967-68 to 1977-78	(a) 1949-50 to 1964-65	(b) 1967-68 to 1977-78	(a) 1949-50 to 1964-65	(b) 1967-68 to 1977-78
Crop						
Foodgrains	2.98	2.40	1.34	0.38	1.61	1.53
Non-Food	3.65	2.70	2.52	1.01	1.06	1.15
All Crops	3.20	2.50	1.60	0.55	1.60	1.40
Rice	3.37	2.21	1.26	0.74	2.09	1.46
Wheat	3.07	5.73	2.70	3.10	1.24	2.53
Pulses	1.62	0.20	1.87	0.75	0.24	0.42

a) Gleaned from NCAR 1976 (vol 1, ch. 3, p. 230-241)
b) Estimates of Area and Production of Principal Crops in India, 1978-79; published by the Directorate of Economic Statistics.

large task. Each aim must be held separately and pursued by separate action. Where there are complementarities they should be exploited. But conflict in programme content must be solved quickly at the political level with a full recognition that if the pursuit of production is made subordinate to these aims, the dismal record of the past will not be altered.'[26]

The record of the achievements of increased production through distributive justice is available in the experience of both Mexico and India in the years prior to the Green Revolution.

Gustava Esteva reports how as a result of the land reforms of the 1930s, the 'ejidos' or lands returned to peasant communities accounted for more than half of the total arable land of the country, and by 1940, for 51% of the total agricultural production. The production of the period continuously expanded at an annual rate of 5.2% from 1935 to 1942.

Similarly, Jatindar Bajaj in his study of pre-and post Green Revolution performance shows that the rate of growth of aggregate crop production was higher in the years before the Green Revolution than after it. 1967-68 is the year the Green Revolution was officially launched in India. (Table 1.4)

The record of agricultural production before the Green Revolution was clearly not 'dismal'. Nor has the record of production been miraculous after the introduction of the 'miracle' seeds. The usual image that is created to support the image of the 'miracle' is that India was transformed from 'the begging bowl to a bread basket'[27] by the Green Revolution and food surpluses put an end to India's living in a 'ship-to-mouth' existence. This common belief is based on the impression that foodgrain imports after the Green Revolution substantially declined. In fact, however, food imports have continued to be significant even after the Green Revolution as illustrated in Table 1.5.

A second reason for the Green Revolution being seen as a miracle lies in an ahistorical view of grain trade. The flow of grain from North to South is of recent origin, (Figure 1) before which, grain travelled from the South to the North. India was a major supplier of wheat to Europe until the war years. As Dan Morgan reports,

> 'In 1873, with the opening of the Suez Canal, the first wheat arrived from India, after a push by British entre-preneurs to obtain a cheap, secure source of wheat under British control. The British envisaged India as a poten-tially secure source of wheat for the Empire. Industrial tycoons pushed rail roads and canals into the Indus and

Table 1.5: Imports of foodgrains in India on Government of India account

Year	Quantity in thousand tonnes
1949	3,765
1950	2,159
1951	4,801
1952	3,926
1953	2,035
1954	843
1955	711
1956	1,443
1957	3,646
1958	3,224
1959	3,868
1960	5,137
1961	3,495
1962	3,640
1963	4,556
1964	6,266
1965	7,462
1966	10,058
1967	8,672
1968	5,694
1969	3,872
1970	3,631
1971	2,054
1972	445
1973	3,614
1974	4,874
1975	7,407
1976	6,483
1977	547

Source: *Directorate of Economics and Statistics, New Delhi*

Ganges river basins, where farmers had been growing wheat for centuries.'[28]

According to George Blyn, in the quarter century before World War I, rising per capital output and consumption pervaded all major regions.

'Most foodgrain crops also expanded at substantial rates,

Figure 1

Major World grain routes, 1880

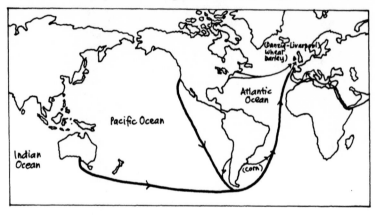

Major World grain routes, 1978

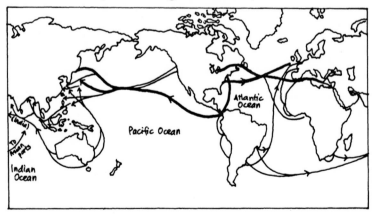

*and though much rice and wheat were exported, domestic availability grew at about the same rate as output...
This early period gives evidence that per capita consumption of agricultural commodities increased over a substantial period of years.'[29]*

In times of crisis and scarcity,the colonial government of course put its revenue needs above those of the survival of the people. On 3 November 1772, a year after the great famine in Bengal that killed about 10 million people, Warren Hastings wrote to the Court of Directors of the East India Company:

'Notwithstanding the loss of at least one third of the inhabitants of the province, and the consequent decrease of the cultivation the net collection of the year 1771 exceeded even those of 1768 ... It was naturally to be expected that the diminution of the revenue should have kept an equal pace with the other consequences of so great a calamity.That it did not was owing to its being violently kept up to its former standard.'[30]

Injustice has been at the root of the worst forms of scarcity throughout human history and injustice and inequality has also been at the root of societal violence. By separating issues of agricultural production from issues of justice, the Green Revolution strategy attempted to diffuse political turmoil. But by-passing the goals of equality and sustainability led to the creation of new inequalities and new scarcities. The Green Revolution strategy for peace had boomeranged. In creating new polarisation, it created new potential for conflict.

As Binswager and Futten noted:

'It does seem clear, however, that the contribution of the

*new seed fertilizer technology to food grain production
has weakened the potential for revolutionary change in
political and economic institutions in rural areas in many
countries in Asia and in other parts of the developing
world. In spite of widening income differentials,the gains
in productivity growth, in those areas where the new
seed-fertilizer technology has been effective, have been
sufficiently diffused to preserve the vested interests of
most classes in an evolutionary rather than a revolutionary pattern of rural development.*

*'By the mid-1970s, however, the productivity gains that
had been achieved during the previous decade were
coming more slowly and with greater difficulty in many
areas. Perhaps revolutionary changes in rural institutions that the radical critics of the Green Revolution for
the past ten years have been predicting will occur as a
result of increasing immiserization in the rural areas of
many developing countries during the coming decade.'*[31]

References 1

1.   Jack Doyle, *Altered Harvest*, New York: Viking, 1985, p256.

2.   S Harding, *The Science Question in Feminism*, Ithaca: Cornell Uni-
      versity Press, p30.

3.   Vandana Shiva, 'Reductionist Science as Epistemic Violence', in A
      Nandy (ed), *Science, Hegemony and Violence*, United Nations Uni-
      versity, Delhi: Oxford University Press, 1988.

4.   Alfred Howard, *The Agricultural Testament*, London: Oxford Uni-
      versity Press, 1940.

5.   John Augustus Voeleker, 'Report on the Improvement of Indian
      Agriculture', London: Eyre and Spothswoode, 1893, p11.

6.   M K Gandhi, *Food Shortage and Agriculture*, Ahmedabad: Najivan
      Publishing House, 1949, p47.

7. K M Munshi, *Towards Land Transformation*, Ministry of Food and Agriculture, undated, p145.

8. C Subramaniam, *The New Strategy in Agriculture*, New Delhi: Vikas, 1979.

9. Jaganath Pathy, 'Green Revolution in India', paper presented at seminar on 'The Crisis in Agriculture', APPEN/TWN, Penang, January 1990.

10. E Taboada, quoted by Gustavo Esteva in 'Beyond the Knowledge/Power Syndrome: The Case of the Green Revolution', paper presented at UNU/WIDER Seminar, Karachi, January 1989, p19.

11. E Hyam, *Soil and Civilisation*, London: Thames and Hudson, 1952.

12. A S Johnson, 'The Foundations Involvement in Intensive Agricultural Development in India', in *Cropping Patterns in India*, New Delhi: ICAR, 1978.

13. Jack Doyle, *op cit*, p256.

14. Claude Alvares, 'The Great Gene Robbery', *Illustrated Weekly of India*, 23 March, 1986.

15. B Onate, 'Why the Green Revolution has failed the small farmers', paper presented at CAP seminar on Problems and Prospects of Rural Malaysia', Penang, November 1985.

16. Frances Moore Lappe and Joseph Collins, *Food First*, London: Abacus, 1982, p114.

17. Augus Wright, 'Innocents Abroad: American Agricultual Research in Mexico', in Wes Jackson, *et al* (ed), *Meeting the Expectations of the Land*, San Francisco: North Point Press, 1984.

18. R P Dutt, quoted in J Bajaj, 'Green Revolution: A Historical Perspective', paper presented at CAP/TWN Seminar on 'The Crisis in Modern Science', Penang, November 1986, p4.

19. J Bajaj, *op cit*, p4.

20. G Esteva, *op cit*, p19.

21. Lappe and Collins, *op cit*, p114.

22. Robert Anderson and Baker Morrison, *Science, Politics and the Agricultural Revolution in Asia*, Boulder: Westview Press, 1982, p7.

23. Anderson and Morrison, *op cit*, p5

24. Anderson and Morrison, *op cit*, p3.

25. Harry Cleaver, 'Technology as Political Weaponry', in Anderson, et al, *Science, Politics and the Agricultural Revolution in Asia*, Boulder: Westview Press, 1982, p269.

26. David Hopper, quoted in Andrew Pearse, *Seeds of Plenty, Seeds of Want*, Oxford: Oxford University Press, 1980, p79.

27. M S Swaminathan, *Science and the Conquest of Hunger*, Delhi: Concept, 1983, p409.

28. Dan Morgan, *Merchants of Grain*, New York: Viking, 1979, p36.

29. George Blyn, 'India's Crop output Trends; Past and Present', C M Shah, (ed), *Agricultural Development of India, Policy and Problems*, Delhi: Orient Longman, 1979, p583.

30. Quoted in J Bajaj, *op cit*, p5.

31. Quoted in Edmund Oasa, 'The political economy of international agricultural research: a review of the CGIAR's response to criticisms of the Green Revolution', in B Gleaser, (ed), *The Green Revolution Revisited*, Boston: Allen and Unwin, 1956, p25.

2

'MIRACLE SEEDS' AND THE DESTRUCTION OF GENETIC DIVERSITY

IN THE 1950s, when Borlaug created the semi-dwarf high yielding variety of wheat, a new religion was born – the religion of the Green Revolution which – promised abundance through the 'miracle seeds'. In 1960, when Borlaug addressed scientists and United Nations officials in Rome, he proposed setting up a programme in Mexico for training agronomists from across the world. True to the religious mould, he called it the 'Practical school of wheat apostles'. The Rockefeller Foundation gave the finances, the FAO, its stamp of intergovernmental legitimisation, and the Mexican government, its facilities.

> 'Borlaug's apostles – who initially came from countries such as Afghanistan, Cyprus, Egypt, Ethiopia, Iran, Iraq Jordan, Libya, Pakistan, Syria, Saudi Arabia, and more than ten countries in South America – were trained by CIMMYT scientists in genetics, agronomy, soils, and plant breeding for one year, and then sent back to their native lands to preach the new agricultural gospel.' [1]

In 1963, Dr M S Swaminathan, one of Borlaug's wheat

apostles, arranged for the high priest to visit India and
spread his gospel of the 'miracle' seeds. Borlaug's visit led
him to believe that,

> 'There is the prospect of a spectacular breakthrough in
> grain production for India. The new Mexican varieties that
> I have seen growing incline me to say that they will do well
> in India and grow beautifully. If the disease resistance
> holds up in its present spectrum.... and if the quantities of
> fertilizer and other vital necessities are made available
> then there is a chance that something big could happen.'[2]

By the mid-1960s, India agricultural policies were ad-
justed to utilise and promote the new seeds. The programme
came to be known as the New Agricultural Strategy. It
concentrated on one tenth of the cultivable land, and initially
on only one crop-wheat. By the summer of 1965, India with
Pakistan, had ordered 600 tonnes of wheat seed from Mexico. In the fall of 1966, India spent $2.5 million for 18,000 tons
of Mexican wheat seed. By 1968, nearly half the wheat
planted came from Borlaug's dwarf varieties. The gospel
spread so fast that by 1972/73, 16.8 million hectares were
planted with dwarf wheat and 15.7 million hectares were
planted with dwarf rice across the Third world. 94% of the
hybrid rice area was in Asia of which nearly half was in India.

The dwarf gene was essential to the technological package of the Green Revolution, which was based on intensive
inputs of chemical fertilizers. The taller traditional varieties
tended to 'lodge' with high applications of chemical fertilizers because they converted the nutrients into overall plant
growth. The shorter, stiffer stems of dwarf varieties allowed
more efficient conversion of fertilizer into grain. The dwarf
genes for wheat were derived from a Japanese variety called
Norin 10, and the dwarf genes for rice came from a Tai-
wanese variety called Dee-Geo-Woo-Gen. The linkage be-

tween chemical fertilizers and dwarf varieties that were established through the breeding programmes of CIMMYT and IRRI, created a major shift in how seeds were perceived and produced, and who controlled the production and use of seeds.

For 10,000 years, farmers and peasants had produced their own seeds, on their own land, selecting the best seeds, storing them, replanting them, and letting nature take its course in the renewal and enrichment of life. With the Green Revolution, peasants were no longer to be custodians of the common genetic heritage through the storage and preservation of grain. The 'miracle seeds' of the Green Revolution transformed this common genetic heritage into private property, protected by patents and intellectual property rights. Peasants as plant breeding specialists gave way to scientists of multinational seed companies and international research institutions like CIMMYT and IRRI. Plant breeding strategies of maintaining and enriching genetic diversity and self-renewability of crops were substituted by new breeding strategies of uniformity and non-renewability, aimed primarily at increasing transnational profits and First World control over the genetic resources of the Third World. The Green Revolution changed the 10,000-year evolutionary history of crops by changing the fundamental nature and meaning of 'seeds'.

For 10,000 years, agriculture has been based on the strategy of conserving and enhancing genetic diversity.

According to former FAO genetic resources expert Erna Bennet,

> *'The patchwork of cultivation sown by man unleashed an explosion of literally inestimable numbers of new races of cultivated plants and their relatives. The inhabited*

*earth was the stage for 10,000 years, for an unrepeatable
plant breeding experiment of enormous dimensions.'³*

In this experiment, millions of peasants and farmers
participated over thousands of years in the development
and maintenance of genetic diversity. The experiment was
concentrated in the so-called developing world of where the
greatest concentrations genetic diversity are found, and
where humans have cultivated crops the longest. The traditional breeders, the Third World peasants, as custodians of
the planet's genetic wealth, treated seeds as sacred, as the
critical element in the great chain of being. Seed was not
bought and sold, it was exchanged as a free gift of nature.
Throughout India, even in years of scarcity, seed was conserved in every household, so that the cycle of food production was not interrupted by loss of seeds.

The shift from indigenous varieties of seeds to the Green
Revolution varieties involved a shift from a farming system
controlled by peasants to one controlled by agrichemical and
seed corporations, and international agricultural research
centers. The shift also implied that from being a free resource
reproduced on the farm, seeds were transformed into a
costly input to be purchased. Countries had to take international loans to diffuse the new seeds, and farmers had to
take credit from banks to use them. International agricultural centers supplied seeds which were then reproduced,
crossed and multiplied at the national level.

For the production of seeds for chemically intensive
agriculture, seeds are classified into four categories by seed
certifying agencies :

1. *Breeder seeds* – Seeds or vegetative propagating materials
   directly produced or controlled by the originating plant
   breeder or institution. Breeder seeds are also called
   nucleus seeds.

Steps For The Production Of Different Seeds

Breeder's seeds   Produced at breeder's institution

Variety | Released

Foundation seeds   Increased from breeder seeds through
a Foundation seeds Programme at
National Seeds Corporation.

Variety | Distributed

Registered seeds   Increased from foundation seeds or
other registered seeds by private seed
growers seed companies.

This step is not very essential

Certified seeds   Increased from foundation/registered
or other certified seeds by private seed
growers and seed companies.

Sold to the farmers

2. *Foundation seeds* – These are the direct increase of breeder seeds.
3. *Registered seeds* are the progeny of foundation seeds.
4. *Certified seeds* are the progeny of foundation, or registered seeds.

World Bank finances were an important element in the spread of the vast network that was needed for distribution of Green Revolution varieties. In 1963, the National Seed Corporation was established. In 1969, the Terai Seed Corporation was started with a World Bank loan of US$13 million. This was followed by two National Seeds Project (NSP) loans. NSP I of US$25 million was given in 1976 and NSP II of US $ 16 million was given in 1978 to support the National Seed Program. The overall objective of the projects was to develop state institutions and create a new infrastructure for increasing the production of certified seeds. In 1988, the World Bank gave India a fourth loan for the seed sector to make India's seed industry more 'market responsive'.

The involvement of the private sector, including multinational corporations, in seed production is a special objective of NSP III (US$150 million). This was viewed as necessary because as the project document notes, 'sustained demand for seeds did not expand as expected, constraining the development of the fledgling industry. In the self-pollinated crops, especially wheat and rice, farmer retention and farmer to farmer transfer accounted for much of the seed used, while some of the HYV's were inferior in grain quality to traditional types and thus lost favour among farmers.'[4] The growth of marketed seeds is thus the main objective of 'developing' the seed 'industry', because farmers' own seeds do not generate growth in financial terms.

The fact that inspite of miracle seeds, farmers in large parts of India prefer to retain and exchange seeds among themselves, outside the market framework, is not taken as an indicator of better viability of their own production and exchange network. It is instead viewed as reason for a bigger push for commercialisation, with bigger loans and better incentives to corporate producers and suppliers. The existence of the indigenous seed industry as a decentralised community based activity is totally eclipsed in the World Bank perspective according to which , 'before the 1960s, the seed industry was little developed'.

The commercialisation of seeds of staple food grains was a contribution from the Green Revolution.

As Jack Doyle comments, 'The first recognition that there was money to be made in the Green Revolution came from the value placed on the seed itself'. With the Green Revolution, seeds were not merely converted into money, they were also converted into machines.

A text book on high-yielding crop varieties states:

'Plants are the primary factory of agriculture where seeds are like the "machine"; fertilizers and water are like the fuel; herbicides, pesticides, equipments, credits and technical know-how are accelerators, to increase the output of this industry. The output in the plant industry is directly correlated with the genetic potential of the seeds to make use of the cash and non-cash inputs. The recent technological revolution in the agriculture industry is described as "as euphemism for the high-yielding varieties of seeds" and their emergence is itself a miracle and a challenge to the Malthusian theory of population and food growth. Inputs like fertilizers, pesticides, herbicides, water, machines and tools have been in use in agriculture since several decades but their utilization was limited to a very low level due to a genetic ceiling on the nutrient uptake in the land varieties.'[5]

The mechanistic thinking underlying the new seed industry is anthromorphic and also culturally chauvinistic. The indigenous varieties, or land races, evolved through both natural and human selection, and produced and used by Third World farmers worldwide are called 'primitive cultivars'. Those varieties created by modern plant breeders in international agricultural research centers or by transnational seed corporations are called 'advanced' or 'elite'.

How the Green Revolution makes Unfair Comparisons

Were the 'miracle' seeds or the Green Revolution inherently superior and 'advanced' in comparison to the diversity of indigenous crops and varieties that they displaced? The miracle of the new seeds has most often been communicated through the term 'high-yielding varieties' (HYVs). The HYV category is a central category of the Green Revolution paradigm. Unlike what the term suggests, there is no neutral or objective measure of 'yield' on the basis of which the cropping systems based on miracle seeds can be established to be higher yielding than the cropping systems they replace. It is now commonly accepted that even in the most rigorous of scientific disciplines such as physics, there are no neutral observational terms. All terms are theory laden.

The HYV category is similarly not a neutral observational concept. Its meaning and measure is determined by the theory and paradigm of the Green Revolution. And this meaning cannot easily and directly be translated for comparison with the agricultural concepts of indigenous farming systems for a number of reasons. The HYV category is essentially a reductionist category which decontextualises contextual properties of both the native and the new varieties. Through the process of decontextualisation, costs and impacts are externalised and systemic comparison with alternatives is precluded.

Cropping systems, in general, involve an interaction

between soil, water and plant genetic resources. In indigenous agriculture, for example, cropping systems include a symbiotic relationship between soil, water, farm animals and plants. Green Revolution agriculture replaces this integration at the level of the farm with the integration of inputs such as seeds and chemicals. The seed/chemical package sets up its own interactions with soils and water systems, which are, however, not taken into account in the assessment of yields.

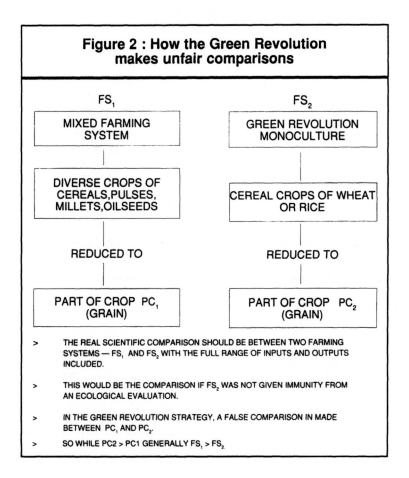

Figure 2 : How the Green Revolution makes unfair comparisons

FS₁ → MIXED FARMING SYSTEM → DIVERSE CROPS OF CEREALS, PULSES, MILLETS, OILSEEDS → REDUCED TO → PART OF CROP PC₁ (GRAIN)

FS₂ → GREEN REVOLUTION MONOCULTURE → CEREAL CROPS OF WHEAT OR RICE → REDUCED TO → PART OF CROP PC₂ (GRAIN)

> THE REAL SCIENTIFIC COMPARISON SHOULD BE BETWEEN TWO FARMING SYSTEMS — FS₁ AND FS₂ WITH THE FULL RANGE OF INPUTS AND OUTPUTS INCLUDED.

> THIS WOULD BE THE COMPARISON IF FS₂ WAS NOT GIVEN IMMUNITY FROM AN ECOLOGICAL EVALUATION.

> IN THE GREEN REVOLUTION STRATEGY, A FALSE COMPARISON IN MADE BETWEEN PC₁ AND PC₂.

> SO WHILE PC2 > PC1 GENERALLY FS₁ > FS₂

Modern plant breeding concepts like HYVs reduce farming systems to individual crops and parts of crops (Figure 2). Crop components of one system are then measured with crop components of another. Since the Green Revolution strategy is aimed at increasing the output of a single component of a farm, at the cost of decreasing other components and increasing external inputs, such a partial comparison is by definition, biased to make the new varieties 'high yielding' even when at the systems level, they may not be. Traditional farming systems are based on mixed and rotational cropping systems, of cereals, pulses, oilseeds with diverse varieties of each crop, while the Green Revolution package is based on genetically uniform monocultures. No realistic assessments are ever made of the yield of the diverse crop outputs in the mixed and rotational systems. Usually the yield of a single crop like wheat or maize is singled out and compared to yields of new varieties. Even if the yields of all the crops were included, it is difficult to convert a measure of pulse into an equivalent measure of wheat, for example, because in the diet and in the ecosystem, they have distinctive functions. The protein value of pulses and the calorie

Table 2.1
Nutritional Content of Different Food Crops
(All values per 100 gms of edible portion)

	Protein (gms)	Minerals (100 gms)	Ca (mg)	Fe (100 gms)
Bajra	11.6	2.3	42	5.0
Ragi	7.3	2.7	344	6.4
Jowar	10.4	1.6	25	5.8
Wheat	11.8	0.6	23	2.5
Rice	6.8	0.6	10	3.1
Bengal Gram	17.1	3.6	202	10.2
Green Gram	24.0	3.5	124	7.3
Rajmah	22.9	3.2	260	5.8

Source: *National Institute of Nutrition, Hyderabad, India.*

value of cereals are both essential for a balanced diet, but in different ways and one cannot replace the other as illustrated in Table 2.1. Similarly, the nitrogen fixing capacity of pulses is an invisible ecological contribution to the yield of associated cereals. The complex and diverse cropping systems based on indigenous varieties are therefore not easy to compare to the simplified monocultures of HYV seeds. Such a comparison has to involve entire systems and cannot be reduced to a comparison of a fragment of the farm system. In traditional farming systems, production has also involved maintaining the conditions of productivity.

The measurement of yields and productivity in the Green Revolution paradigm is divorced from seeing how the processes of increasing output affect the processes that sustain the condition for agricultural production. While these reductionist categories of yield and productivity allow a higher measurement of yields, they exclude the measurement of the ecological destruction that affects future yields. They also exclude the perception of how the two systems differ dra-

Figure 3: Internal Input Farming System

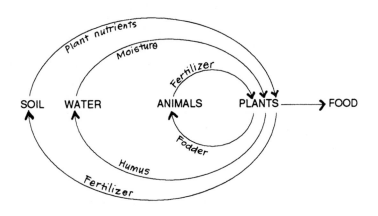

matically in terms of inputs (Figure 3). The indigenous cropping systems are based only on internal organic inputs.
Seeds come from the farm, soil fertility comes from the farm
and pest control is built into the crop mixtures. In the Green
Revolution package, yields are intimately tied to purchased
inputs of seeds, chemical fertilizers, pesticides, and petroleum and to intensive and accurate irrigation. High yields
are not intrinsic to the seeds, but are a function of the
availability of required inputs, which in turn have ecologically destructive impacts (Figure 4).

The Myth of the High Yielding Variety

As Dr Palmer concluded in the United Nations Research
Institute for Social Development's 15 nation study of the
impact of the seeds, the term 'High Yielding Varieties' is a
misnomer because it implies that the new seeds are high-
yielding in and of themselves. The distinguishing feature of
the seeds, however, is that they are highly responsive to
certain key inputs such as fertilizers and irrigation. Palmer
therefore suggested the term 'high-responsive varieties'
(HRV's) in place of 'high yielding varieties'(HYV).[6] In the
absence of additional inputs of fertilizers and irrigation, the
new seeds perform worse than indigenous varieties. With
the additional inputs, the gain in output is insignificant
compared to the increase in inputs. The measurement of
output is also biased by restricting it to the marketable part
of crops. However, in a country like India, crops have traditionally been bred and cultivated to produce not just food
for man but fodder for animals, and organic fertilizer for
soils.

According to A K Yegna Narayan Aiyer, a leading authority on agriculture, 'as an important fodder for cattle and
in fact as the sole fodder in many tracts, the quantity of straw

Figure 4 : External Input Farming System

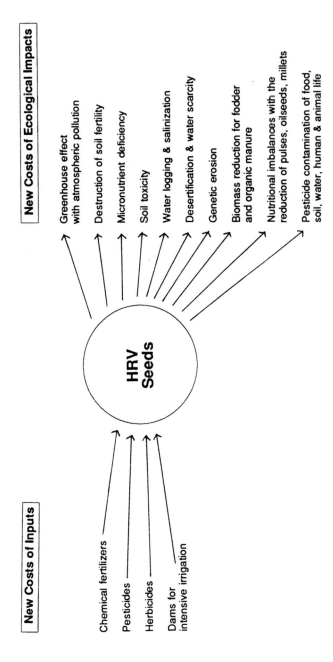

New Costs of Inputs

Chemical fertilizers
Pesticides
Herbicides
Dams for intensive irrigation

HRV Seeds

New Costs of Ecological Impacts

Greenhouse effect with atmospheric pollution
Destruction of soil fertility
Micronutrient deficiency
Soil toxicity
Water logging & salinization
Desertification & water scarcity
Genetic erosion
Biomass reduction for fodder and organic manure
Nutritional imbalances with the reduction of pulses, oilseeds, millets
Pesticide contamination of food, soil, water, human & animal life

obtainable per acre is important in this country. Some varieties which are good yielders of grains suffer from the drawback of being low in respect to straw.[7] He illustrated the variation in the grain-straw ratio of rice with yields from the Hebbal farm. (Table 2.2)

Table 2.2
Grain and Straw Production of Rice Varieties

Name of Variety	Grain (in lb.per acre)	Straw (in lb. per acre)
Chintamani sanna	1,663	3,333
Budume	1,820	2,430
Halubbalu	1,700	2,740
Gidda Byra	1,595	2,850
Chandragutti	2,424	3,580
Putta Bhatta	1,695	3,120
Kavada Bhatta	2,150	2,940
Garike Sanna	2,065	2,300
Alur sanna	1,220	3,580
Bangarkaddi	1,420	1,760
Banku (rainy season 1925-26)	1,540	1,700
G.E.B. -Do-	1,900	1,540

Under the Green Revolution, multiple uses of plant biomass seem to have been consciously sacrificed for a single use, through non-sustainable consumption of fertilizer and water. The increase in marketable output of grain has been achieved at the cost of decrease of biomass for animals and soils and the decrease of ecosystem productivity due to over-use of resources. The increase in production of grain for the market was achieved by reducing the biomass for internal use on the farm.

This is explicit in a statement by Swaminathan:

'High yielding varieties of wheat and rice are high yielding because they can use efficiently larger quantities of nutrients and water than the earlier strains, which tended to

*lodge or fall down if grown in soils with good fertility.... They thus have a "harvest index" (i.e. the ratio of the economic yield to the total biological yield) **which is more favourable to man**. In other words, if a high yielding strain and an earlier tall variety of wheat both produce, under a given set of conditions, 1000 kg of dry matter, the high yielding strain may partition this dry matter into 500 kg for grain and 500 kg for straw. The tall variety, on the other hand, may divert 300 kg for grain and 700 kg for straw.'[8]*

The reduction of output of biomass for straw production was probably not considered a serious cost since chemical fertilizers were viewed as a total substitute for organic manure, and mechanisation was viewed as a substitute to animal power.

According to one author,

'It is believed that the "Green Revolution" type of technological change permits higher grain production by changing the grain foliage ratio.... At a time there is urgency for increasing grain production, an engineering approach to altering the product mix on an individual plant may be advisable, even inevitable. This may be considered another type of survival technological change. It uses more resources, returns to which are perhaps unchanged (if not diminished).'[9]

It was thus recognised that in terms of over-all plant biomass, the Green Revolution varieties could even reduce the overall yields of crops and create scarcity in terms of output such as fodder.

Finally, there is now increasing evidence that indigenous varieties could also be high yielding, given the required inputs. Richaria has made a significant contribution to the recognition that peasants have been breeding high yielding varieties over centuries.

Richaria reports :

'A recent varietal-cum-agronomic survey has shown that nearly 9% of the total varieties grown in U P fall under the category of high yielding types (3,705 kgs and above per hectare.'

'A farmer planting a rice variety called Mokdo of Bastar who adopted his own cultivation practices obtained about 3,700 to 4,700 kgs of paddy per hectare. Another rice grower of Dhamtari block (Raipur) with just an hectare of rice land, falling not in an uncommon category of farmers, told me that he obtains about 4,400 kgs of paddy per hectare from Chinnar variety, a renowned scented type, year after year with little fluctuations, he used FYM supplemented at times with a low dose of nitrogen fertilizers. For low lying areas in Farasgaon Block (Bastar) a non-lodging tall rice variety Surja with bold grains and mildly scented rice may compete with Jaya in yield potential at lower doses of fertilization, according to a local grower who showed me his crop of Surja recently.

'During my recent visit of the Bastar area in the middle of November, 1975 when the harvesting of new rice crop was in full swing in a locality, in one of the holdings of an adivasi cultivator, Baldeo of Bhatra tribe in village Dhikonga of Jugalpur block, I observed a field of Assam Chudi ready for harvest with which the adivasi cultivator has stood for crop competition. The cultivator has applied the fertilizer approximately equal to 50 kg/N ha and has used no plant protection measures. He expected a yield of about 5,000 kg/ha. These are good cases of applications of

an intermediate technology for increasing rice production. The yields obtained by those farmers fall in or above the minimum limits set for high yields and these methods of cultivation deserve full attention.'[10]

India is a Vavilov centre or centre of genetic diversity of rice. Out of this amazing diversity, Indian peasants and tribals have selected and improved many indigenous high yielding varieties. In South India, in semi-arid tracts of the Deccan, yields went up to 5,000 kg/ha under tank and well irrigation. Under intensive manuring, they could go even higher.

As Yegna Narayan Aiyer reports:

'The possibility of obtaining phenomenal and almost unbelievably high yields of paddy in India has been established as the result of the crop competitions organized by the Central Government and conducted in all states. Thus even the lowest yield in these competitions has been about 5,300 lbs/acre, 6,200 lbs/acre in West Bengal, 6,100, 7,950, and 8,258 lbs/acre in Thirunelveli, 6,368 and 7,666 kg/ha in South Arcot, 11,000 lbs/acre in Coorg and 12,000 lbs/ acre in Salem.'[11]

It has often been argued that the Green Revolution strategy was the only available one for increasing food availability. International agencies and Third World governments had no option, we are told. The inevitability of the Green Revolution option was built on neglecting the other avenues for increasing food production that is more ecological, such as improving mixed cropping systems, improving indigenous seeds and improving the efficiency of the use of local resources. Geertz has called this process of the organic intensification of agriculture 'involution'.[12] In contrast to the chemical intensification strategy of the Green Revolution, involution offered higher yields with sustainability, not higher yields at the cost of sustainability. If one further recognises that sustainability involves sustainable livelihoods, not just sustainable output, 'involution' was also a more efficient policy for utilising the labour available in high population regions than the policy of Green Revolution or industrial agriculture.

Comparative studies of 22 rice-growing systems have shown that indigenous systems were more efficient in terms of yields, and in terms of labour use and energy use (Figure 5).

Genetic Uniformity and the Creation of New Pests

Sophisticated breeding strategies were prevalent in agricultural societies long before the Green Revolution.

Dan Morgan, in his study of the global grain trade reports that:

> 'peasants in British India unknowingly made a major contribution to the prosperity of North America. They had patiently developed a wheat that could ripen quickly between Indian monsoons. The seeds developed by these poor peasants subsequently were used in adapting Marquis wheat, which became Canada's premium grain, to the short, northern growing season of the Prairies.'[13]

Diversity was a central principle of these breeding strategies. Diversity contributed to ecological stability, and hence to ecosystem productivity. The less the diversity and the more the uniformity in an ecosystem, the higher is its vulnerability to instability, breakdown and collapse.

Figure 5 : Yield in relation to labour input based on comparison of 22 rice-growing systems

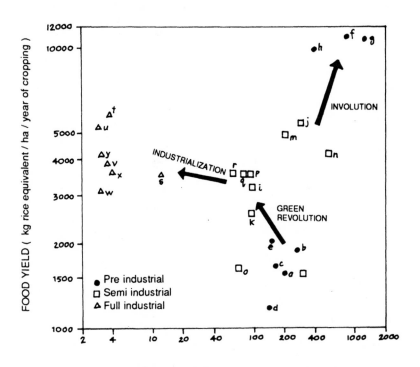

LABOUR INPUT (days / ha / year of cropping)

Source : *Bayliss-Smith, 1980*

Table 2.3
Pre-industrial, semi-industrial and
full-industrial systems of rice cultivation:
inputs and outputs per hectare-year

Location	Fossil fuel input	Labour per crop (days)	Labour as% total input	Total input (GJ)	Total output (GJ)
1	2	3	4	5	6
Pre-industrial					
a.Dayak,Sarawak (1951)	2%	208	44%	0.30	2.4
b.Dayak,Sarawak (1951)	2%	271	51%	0.63	5.7
c.Kilombero, Tansania (1967)	2%	170	39%	0.42	3.8
d.Kilombero, Tansania (1967)	3%	144	35%	1.44	9.9
e.Iban, Sarawak (1951)	3%	148	36%	0.27	3.1
f.Luts'un, Yunnan (1938)	3%	882	70%	8.04	166.9
g.Yits'un, Yunnan (1938)	2%	1293	78%	10.66	163.3
h.Yuts'un, Yunnan (1938)	4%	426	53%	5.12	149.3
Semi-industrial					
i.Mandya, Karnataka (1955)	23%	309	46%	3.33	23.8
j.Mandya, Karnataka (1975)	74%	317	16%	16.73	80.0
k.Phillippines (1972)	86%	102	5.3%	12.37	39.9
l.Philippnes (1972)	89%	102	4.1%	16.01	51.6
m.Japan (1963)	90%	216	5.2%	30.04	73.7
n.Hongkong (1971)	83%	566	12%	31.27	64.8
o.Philippines (1965)	93%	72	13%	3.61	25.0
p.Philippines (1979)	33%	92	16%	5.48	52.9

1	2	3	4	5	6
q.Philippines (1979)	80%	84	11%	6.90	52.9
r.Philippines (1979)	86%	68	7%	8.72	52.9
Full-industrial					
s.Surinam (1972)	95%	12.6	0.2%	45.9	53.7
t.USA (1974)	95%	3.8	0.02%	70.2	88.2
u.Sacramento Calif.(1977)	95%	3.0	0.04%	45.9	80.5
v.Grand prairie Ark.(1977)	95%	3.7	0.04%	52.5	58.6
w.Southwest Louisiana (1977)	95%	3.1	0.04%	48.0	50.8
x.Mississippi Delta (1977)	95%	3.9	0.05%	53.8	55.4
y.Texas Gulf Coast(1977)	95%	3.1	0.04%	55.1	74.7

The Green Revolution package was built on the displacement of genetic diversity at two levels. Firstly, mixtures and rotation of diverse crops like wheat, maize, millets, pulses, and oil seeds were replaced by monocultures of wheat and rice. Secondly, the introduced wheat and rice varieties reproduced over large-scale as monocultures came from a very narrow genetic base, compared to the high genetic variability in the populations of traditional wheat or rice plants. When 'HYV' seeds replace native copying systems diversity is lost irreversibly. The destruction of diversity and the creation of uniformity simultaneously involves the destruction of stability and the creation of vulnerability.

As in the rest of India, indigenous agriculture in Punjab was based on diversity. Among the non-food crops indigo, sugarcane, cotton, hemp, asssafoetida and oilseeds were grown. The horticultural crops included guavas, dates, mangoes, limes, lemons, peaches, apricots, figs, pomegranates, plums, oranges, mulberries, grapes, almonds, melons, apples, beans, cucumbers, carrots and turnips. The uncultivated areas were covered by date-palm, wild palm, willows acacias, sissoo, byr apple etc.

The millets, called 'minor cereals' (because they are so diverse, not because they are insignificant crops), occupied the largest area under cultivation in Punjab. 'Kutki', the little millet (Panicum miliare), 'jawar' (Sorghum vulgare), 'Mandal' or 'Chalodra' (Eleusine Coracana) and 'bajra', or bulrush millet (Pennisetum typhoidenam) were the main millets cultivated in Punjab, covering 43% of the area. Besides these, there were uncultivated or wild varieties of millet like 'shama' (Panicum hydaspicum), cenchrus echinatus, pennisetum cenchroides. In addition to these, were the more well known cereals 'makki' or maize and wheat. Closely related to the millets were lesser known crops like amaranth of which Punjab had a rich diversity, 'Sil' or 'mawal' (Celosia cristata) grew both cultivated or wild. 'Gauhar' 'sawal' sil, bhabri, savalana, batu, chaulei were the different names given to the common amaranth (Amarantus paniculatus). An important crop used for green leafy vegetable was bathua (Chenopodium album). The pulses of Punjab included 'moth-safaid' (Cyamopsis psoralioides), 'channa' or chick pea (Cicer arietinuin), 'bhut' (Glysine soja), 'urd' and 'mash' (Phascolusmungo) 'lobiya' (Phaseolus lunatus) 'rawan' (Vigna catinag) 'Kalat' (Dolixhoa bidloeua) 'Kharnab-mibti' (Ceratonia siliqua). The oilseeds including 'till' or sesame (seasamum indicum), groundnuts (Arachis hypogea), 'alsi' or linseed (Linum usitatissimum) and 'sarson' or mustard (Brassica nigra). The cereals, pulses and oilseeds were grown in various mixtures and rotations.[14]

As a result of the Green Revolution in Punjab, common lands under forests and pastures have been put under agricultural crops.As the Green Revolution spread, local community management broke down and grazing lands and forests were broken up for monoculture cultivation. Today 4.21 million hectares out of 5.03 million hectares, or 84% of the geographical area of the Punjab is under cultivation, compared to only 42% for India as a whole. Only 4% of the Punjab is now under 'forest', most of it being man-made plantations of eucalyptus (Figure 6).[15]

As the marginal lands and croplands are homogenized, diversity disappears. Genetic diversity in Punjab has been destroyed by the Green Revolution at two other levels-first by the transformation of mixed and rotational cropping of wheat, bajra, jowar, barley, pulses and oilseeds into monocultures and multicropping of wheat and rice, and second, by the conversion of wheat and rice from diverse native varieties suited to different soil, water and climatic conditions to monocultures of single varieties derived from the exotic dwarf varieties of CIMMYT and IRRI.

Between 1966-67 and 1985-86, the cropped area in Punjab increased from 51.71 lakh hectares to 71.76 hectares, and the cropping intensity increased from 133.1 to 170%. The percentage of area under cereals increased from 51 to 72-80 % while area under pulses such as masoor, arhar, moong, Bengal gram declined from 13.38 to 3.48%. The maximum decline in area under pulses was witnessed in the case of the gram crop, where its share in total cropped area declined from 12.26% in the base year of 1966-67 to only 1.39% in the year 1985-86. The area under oilseeds like rape and mustard and groundnut declined from 6.24 in 1966-67 to 2.93% in 1985-86. The cropping pattern has thus witnessed a major shift in favour of wheat in the rabi season and paddy in the kharif season. Wheat has spread at the cost of gram, barley, rape and mustard which were usually sown as mixed crops with traditional wheat varieties. Similarly, the area under

Table 2.4 : Areas under Different Crops

Year	Wheat	Paddy	Maize	Cereals	Gram	Arhar	Mash	Moong
1966-67	1608.0	285.0	444.0	2634.0	634.0	1.0	22.0	3.0
	(31.09)	(5.50)	(8.59)	(51.0)	(12.26)	(0.02)	(0.43)	(0.06)
1967-68	1790.0	314.0	476.0	2945.0	530.0	1.0	29.0	4.0
	(32.89)	(5.77)	(8.75)	(54.13)	(9.74)	(0.02)	(0.53)	(0.07)
1968-69	2063.0	345.0	490.0	3186.0	348.0	1.0	29.0	4.0
	(38.50)	(6.52)	(9.27)	(60.25)	(6.58)	(0.02)	(0.53)	(0.08)
1969-70	2166.0	359.0.	534.0	3348.0	380.0	2.0	24.0	3.0
	(39.38)	(6.52)	(9.71)	(60.88)	(6.91)	(0.04)	(0.44)	(0.05)
1970-71	2299.0	390.0	555.0	3514.0	358.0	3.0	26.0	7.0
	(40.48)	(6.86)	(9.77)	(61.89)	(6.31)	(0.05)	(0.46)	(0.12)
1971-72	2336.0	450.0	548.0	3531.0	335.0	2.0	23.0	3.0
	(40.81)	(7.85)	(9.57)	(61.69)	(5.85)	(0.03)	(0.40)	(0.05)
1972-73	2404.0	476.0	562.0	3634.0	319.0	2.00	31.0	5.0
	(40.53)	(8.02)	(9.48)	(61.27)	(5.38)	(0.03)	(0.52)	(0.08)
1973-74	2338.0	499.0	567.0	3668.0	352.0	5.0	39.0	10.0
	(38.72)	(8.26)	(9.39)	(60.76)	(5.83)	(0.08)	(0.65)	(0.17)
1974-75	2207.0	569.0	522.0	3627.0	266.0	3.0	27.0	6.0
	(37.38)	(9.63)	(8.84)	(61.43)	(4.51)	(0.05)	(0.46)	(0.10)
1975-76	2439.0	567.0	577.0	3891.0	381.0	6.0	24.0	5.0
	(38.99)	(9.06)	(9.22)	(62.21)	(6.09)	(0.10)	(0.38)	(0.08)
1976-77	2630.0	680.0	536.0	4066.0	352.0	4.03	18.0	4.0
	(41.84)	(10.81)	(8.53)	(64.69)	(5.55)	(0.06)	(0.29)	(0.06)
1977-78	2620.0	856.0	445.0	4077.0	352.0	5.0	19.0	4.0
	(41.0)	(13.39)	(6.96)	(63.80)	(5.52)	(0.08)	(0.30)	(0.10)
1978-79	2738.0	1053.0	424.0	4345.0	350.0	18.0	22.0	3.00
	(41.29)	(15.88)	(6.40)	(65.54)	(5.29)	(0.12)	(0.33)	(0.05)
1979-80	2813.0	1172.0	393.0	4472.0	236.0	13.0	20.0	5.0
	(43.04)	(17.93)	(6.01)	(68.43)	(3.61)	(0.26)	(0.31)	(0.08)
1980-81	2812.0	1183.0	382.0	4513.0	258.0	18.0	21.0	14.0
	(41.57)	(17.49)	(5.65)	(66.73)	(3.81)	(0.27)	(0.31)	(0.20)
1981-82	2914.0	1269.0	340.0	4674.0	243.0	13.0	18.0	25.0
	(42.05)	(18.31)	(4.91)	(67.46)	(3.510)	(0.19)	(0.26)	(0.36)
1982-83	3051.0	1322.0	308.0	4808.0	124.0	18.0	16.0	29.0
	(44.12)	(19.11)	(4.45)	(69.53)	(1.79)	(0.26)	(0.23)	(0.42)
1983-84	3123.0	1482.0	293.0	5006.0	96.0	40.0	14.0	33.0
	(44.75)	(21.23)	(4.20)	(71.7)	(1.38)	(0.57)	(0.20)	(0.47)
1984-85	3096.0	1644.0	304.0	5151.0	104.0	42.0	12.0	34.0
	(44.15)	(23.44)	(4.33)	(73.4)	(1.48)	(0.60)	(0.17)	(0.48)
1985-86	3150.0	1703.0	260.0	5224.0	100.0	40.0	13.0	45.0
	(43.90)	(23.73)	(3.62)	(72.8)	(1.39)	(0.56)	(0.18)	(0.63)

*The total cropped area for 1985-86 includes area under fruits, vegetables and fodders. Figures in the parentheses indicate percentages to total cropped area.

in Punjab 1966-67 to 1985-86   (000' hectares)

Pulses Mustard	Food	Rape & Oilseeds	G.Nut	Total	S.Cane	Potato	Cotton (Ann+ Desi)	Cotton Am.	Total cropped area
692.0	3326.0	119.0	179.0	323.0	156.0	14.0	435.0	199.0	5171
13.38)	(64.32)	(2.30)	(3.46)	(6.24)	(3.2)	(0.27)	(8.41)	(3.85)	
597.0	3542.0	159.0	222.0	399.0	137.0	17.0	419.0	227.0	5441
(10.97)	(65.10)	(12.92)	(14.08)	(7.33)	(2.52)	(0.31)	(7.70)	(4.17)	
411.0	3597.0	70.0	222.0	307.0	157.0	16.0	392.0	229.0	5288
(7.77)	(68.02)	(1.32)	(4.20)	(5.80)	(2.97)	(0.30)	(7.41)	(4.33)	
433.0	3781.0	92.0	186.0	294.0	149.0	16.0	409.0	221.0	5499
(7.87)	(68.76)	(1.67)	(3.38)	(5.34)	(2.71)	(0.29)	(7.44)	(4.02)	
414.0	3928.0	103.0	1740.0	295.0	128.0	17.0	(397.0)	(212.0)	5678
(7.29)	(69.18)	(1.81)	(3.06)	(5.19)	(2.25)	(0.30)	(6.99)	(3.73)	
384.0	3915.0	128.0	174.0	319.0	103.0	17.0	475.0	246.0	5724
(6.71)	(68.40)	(2.24)	(3.04)	(5.57)	(1.80)	(0.30)	(8.30)	(4.30)	
381.0	4015.0	172.0	160.0	351.0	102.0	16.0	506.0	235.0	5931
(6.42)	(67.70)	(2.90)	(2.70)	(5.92)	(1.72)	(0.27)	8.53)	(3.96)	
431.0	4099.0	179.0	155.0	357.0	110.0	23.0	524.0	301.0	6037
(7.14)	(67.90)	(2.97)	(2.57)	(5.91)	(1.82)	(0.38)	(8.66)	(4.99)	
330.0	3957.0	179.0	164.0	372.0	123.0	20.0	547.0	342.0	5904
(5.55)	(67.02)	(3.03)	(2.78)	(6.30)	(2.08)	(0.34)	(9.26)	(5.79)	
441.0	4332.0	122.0	168.0	315.0	114.0	27.0	580.0	365.0	6255
(7,05)	(69.26)	(1.95)	(2.69)	(5.04)	(1.82)	(0.43)	(9.27)	(5.80)	
395.0	4461.0	67.0	164.0	250.0	113.0	29.0	555.0	375.0	6285
(6.28)	(70.98)	(1.07)	(2.61)	(3.98)	(1.80)	(0.46)	(8.83)	(5.97)	
402.0	4479.0	128.0	156.0	287.0	115.0	37.0	609.0	440.0	6390
(6.29)	(70.09)	(2.0)	(2.44)	(4.49)	(1.80)	(0.58)	(9.63)	(6.89)	
410.0	4775.0	83.0	129.0	230.0	108.0	53.0	631.0	470.0	6630
(6.18)	(71.72)	(1.25)	(1.950)	(3.47)	(1.63)	(0.80)	(9.52)	(7.09)	
290.0	4762.0	88.0	91.0	199.0	77.0	41.0	630.0	460.0	6535
(4.44)	(72.87)	(1.35)	(1.39)	(3.05)	(1.18)	(0.63)	(9.64)	(7.04)	
341.0	4854.0	146.0	83.0	238.0	71.0	40.0	649.0	502.0	6763
(5.04)	(71.77)	(2.16)	(1.23)	(3.520)	(1.05)	(0.59)	(9.60)	(7.42)	
325.0	4999.0	110.0	92.0	225.0	104.0	36.0	686.0	546.0	6929
(4.69)	(72.15)	(1.59)	(1.33)	(3.25)	(1.50)	(0.48)	(9.90)	7.88)	
207.0	5015.0	85.0	78.0	187.0	103.0	30.0	724.0	583.0	6915
(2.99)	(72.52)	(1.23)	(1.13)	(2.70)	(1.49)	(0.33)	(10.47)	8.43)	
200.0	5206.0	78.0	58.0	157.0	84.0	30.0	650.0	556.0	6978
(2.9)	(74.6)	(1.12)	(0.83)	(2.20)	(1.27)	(0.47)	(9.3)	(7.97)	
216.0	5367.0	131.0	45.0	198.0	84.0	34.0	472.0	410.0	7013
(3.17)	(76.5)	(1.87)	(0.64)	(2.87)	(1.27)	(0.5)	(6.7)	(5.85)	
250.0	5474.0	150.0	46.0	210.0	95.0	35.0	558.0	471.0	7169
(3.48)	(76.8)	(2.09)	(0.64)	(2.93)	(1.32)	(0.49)	(7.78)	(6.56)	

Source: *Statistical Abstracts of Punjab and officially released data by the Directorate of Agriculture. Punjab.*

paddy has increased at the cost of maize, kharif pulses like moong and masoor, groundnut, green fodder and cotton. And all the wheat and rice monocultures are derived from the genetically narrow base of the Borlaug wheats and IRRI rice.

Linked to the centralized strategy for breeding 'voracious varieties' of seeds which consume high level of fertilizers and water was the need for uniformity and the destruction of diversity. Uniformity became imperative both from the view of centralized production of seeds as well as centralized provisioning of inputs like water and fertilizers.

The wheat seeds that spread worldwide from CIMMYT through Borlaug and his 'wheat apostles' were the result of nine years of experimenting with Japanese 'Norin' wheat. 'Norin' released in Japan in 1935 was a cross between Japanese dwarf wheat called 'Daruma' and American wheat called 'Fultz' which the Japanese government had imported from the US in 1887. The Norin wheat was brought to US in 1946 by Dr D C Salmon, an agriculturist acting as a US military adviser in Japan, and further crossed with American seeds of the variety called 'Bevor' by USDA scientist Dr Orville Vogel. Vogel in turn sent it to Mexico in the 1950s where it was used by Borlaug, who was on the Rockefeller Foundation staff, to develop his well-known Mexican varieties. Of the thousands of dwarf seeds created by Borlaug, only three went to create the 'Green Revolution' wheat plants which were spread worldwide. On this narrow and alien genetic base, are the food supplies of millions precariously perched.

The All India Coordinated Wheat Improvement Project was started in 1964-65 after Borlaug's visit in 1963 which led to the import of 18,000 tonnes of 150 of dwarf varieties from Mexico. Scientists of the Punjab Agricultural University

Figure 6 : Land Utilization (Punjab State) 1978-79

 Other uncultivated land
excluding fallow land

 Fallow land

 Nett area sown

 Land not available for
cultivation

 Forest area

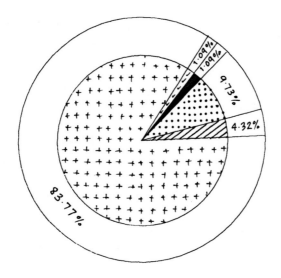

Source : *Statistical abstract of Punjab*

selected two strains from the imported seeds and these were multiplied in the high valleys of Lahaul-Spiti in May-September 1964. PV 18, a cross (Panjino Sb Gabo-55) was the first dwarf variety multiplied. In 1968 two other dwarf varieties,Kalyansona and Sonalika were put out. In the year 1977-78, 100,000 tonnes of certified wheat seed were earmarked for production. Of this, 65,000 tonnes was Sonalika and 30,000 tonnes was Kalyansona variety. Kalyansona was derived from a Mexican cross (Fn-K58 New Thatcher X Norin 10-Brever)X Gabo. Sonalika was derived from a Mexican cross II-54-388-AN X Yt.50 x N10V LR III 8427.[16] Gill has tried to correct the false notion that the Mexican dwarf wheat varieties were the first improved varieties of wheat to be seen by the Punjab peasant.[17] In 1952, a wheat crop in Ludhiana had yielded 5,900 kg/ha under irrigated conditions. Alfred Howard, during his stay at the Imperial research station at Pusa had, with local expertise, produced improved wheats in the first two decades of the present century. By the First World War, over a million acres were planted with the Pusa wheats. By the end of the Second World War more than 2.8 m hectares were under improved varieties, which included type ll bred in 1913 and type 8A bred in 1919 by local peasants and local scientists. In the 1930s, Dr Ram Dhan Singh at the Agricultural College and Research Institute, Lyallpur, bred two varieties C518 and C519 which had a great impact. In 1965, the Punjab Agricultural University released a tall, high yielding wheat variety, C306, with an average yield of 3291 kg/ha, and when measured by per unit water use, or unit fertilizer use, was a higher yield than the Borlaug wheat varieties.

Large scale monocultures of exotic varieties of wheat has turned minor diseases such as Karnal Bunt into epidemic proportions. Leaf Blight Brown rust and Loose Smut are other diseases in wheat. High yielding wheat varieties, viz. PV18, Kalyan 227, Sonara 64 and Leema Rojo are highly susceptible to alternaria leaf blight and glume blotch, and the foot-rot and seedling blight disease caused by Alternaria and Helminthos porium. Unlike the traditional high yielding varieties which have co-evolved with local ecosystems, the Green Revolution HYV's have to be replaced frequently. Seeds, a renewable resource, are thus converted into a non-renewable resource, with each variety usable for only one or two years before it gets overtaken by pests. Obsolescence replaces sustainability. As a text book on high yielding varieties of crops admits, 'The high yielding varieties and hybrids have three to five years life in the field. Thereafter they become susceptible to the new races and biotypes of diseases and pests.'

The vulnerability of rice to new pests and diseases due to monocropping and a narrow genetic base is also very high. In 1966, IRRI released IR-8, an advanced rice variety that came from a cross between and Indonesian variety called 'Peta' and another from Taiwan called 'Dee-Geo-Woo-Gen'. IR-8, Taichung Native 1 and other varieties were brought to India and became the basis of The All India Co-ordinated Rice Improvement Project to evolve dwarf, photo-insensi-tive, short duration, high yielding varieties of rice suited to high fertility conditions. The large scale spread of exotic strains of rice with a narrow genetic base was known to carry the risk of the large-scale spread of diseases and pests. As summarised in a publication titled, 'Rice Research in India– An Overview', from CRRI,

> 'The introduction of high yielding varieties has brought about a marked change in the status of insect pests like gall midge, brown planthopper, leaf folder, whore maggot etc. Most of the high yielding varieties released so far are susceptible to major pests with a crop loss of 30 to 100 %... Most of the HYVs are the derivatives of TN(1) or IR-8 and therefore, have the dwarfing gene of dee-geo-woo-gen. The

narrow genetic base has created alarming uniformity,
causing vulnerability to diseases and pests. Most of the
released varieties are not suitable for typical uplands and
lowlands which together constitute about 75% of the total
rice area of the country.'[18]

Before 1965, rice was an insignificant crop in Punjab.
With the high yielding varieties programme, the area under
rice in Punjab rose dramatically. From 292,000 hectares
producing 292,000 tonnes of rice in 1965, the figures have
shot up to 1,703,000 hectares producing 5,448,000 tonnes.
Table 2.5 shows the increase in area under rice in Punjab.

Table 2.5: Area under rice in the Punjab

Year	Area (000 ha)	Area under HYV (%)
1965	292	-
1966	282	1.0
1967	314	5.4
1968	345	7.5
1969	359	20.0
1970	390	33.3
1971	450	69.1
1972	479	80.5
1973	498	86.8
1974	569	84.5
1975	567	91.2
1976	680	88.4
1977	856	89.5
1978	1052	95.1
1979	1172	91.8
1980	1177	92.6
1981	1270	95.1
1982	1319	94.8
1983	1481	95.0
1984	1645	95.0
1985	1703	95.0

Source: *Sidhu*

The percentage of cropped area under rice increased from 5.5 in 1966-67 to 23.73 in 1985-86. 95% of this area is under semi-dwarf rice varieties, with 5% of the remaining rice area being under Basmati rice.

Table 2.6 lists the semi-dwarf varieties of rice released in Punjab since 1966. New varieties had to be introduced in rapid succession because they were susceptible to new diseases and pests listed in Table 2.7.

Whenever the new IRRI rice varieties were introduced in Asia, they proved to be susceptible to diseases and pests.

Table 2.6: Semi dwarf rice varieties released in Punjab

Year of release	Variety	Cross
1966	Taichung Native 1	DGWG/Tsai Yuan Chung
1968	IR 8	Prta/DGWG
1971	Jaya	T(N) 1/T 141
1972	RP5-3 (Sona)	GEB 24/TN 1
1972	Palman 579	IR 8/Tadukan
1972	RM 95	Jhona349/T(N)1(IrrF2)
1976	PR 106	IR/Peta5/Bella Patna)
1978	PR 103	IR 8/IR 127-2-2
1982	PR 4141	IR 8/PJI/IR 22
1986	PR 108	Vijaya Ptb 21
1986	PR 109	IR 19660-73-4/ IR 21415-90-4/ IR 5853-162-1-2-3

Source: *Sidhu*

IR-8 was attacked by bacterial blight in South East Asia in 1968 and 1969. In 1970 and 1971, it was destroyed by the tungro virus. In 1975, half a million acres under the new rice varieties in Indonesia were destroyed by pests. In 1977, IR-36 was developed to be resistant to 8 major known diseases and pests including bacterial blight and tungro. But this was attacked by two new viruses called 'ragged stunt' and 'wilted stunt'.

Table 2.7: Outbreaks of rice insect pests and diseases in the Punjab

Year	Insect pests/diseases appeared in outbreak form	Varieties	District affected by outbreaks
1967	Leaf-folder	Basmati 370,IR8	Kapurthala,
1972	Root weevil	IR8, Jaya	Patiala
	Whitebacked plant hopper	Sabarmati,Ratna, Palman 579,RP5-3	Ludhiana
1973	Brown plant hopper	IR8, Jaya	Kapurthala,Patiala, Ludhiana, Ropar
1975	Brown plant hopper	IR 8, Jaya	Gurdaspur, Ferozpur,
	Whitebacked plant hopper	IR 8	Kapurthala
1975	Bacterial blight	IR8, Jaya, PR106	Gurdaspur, Amritsar
1978	Whitebacked plant hopper	PR558,PR559, PR562	Kapurthala
	Sheath blight	IR8, Jaya, PR106, PR103	Amritsar, Jalandhar, Kapurthala, Patiala
	Sheath rot	PR 106, IR8	All rice growing areas
1980	Bacterial blight	PR106, IR8, Jaya, PR103, Basmati370	All rice growing areas
	Stem rot	PR106, IR8, Jaya	Amritsar,Gurdaspur, Patiala,Kapurthala, Ferozepur
1981	Whitebacked plant hopper	PR107, PR 4141	Kapurthala,Patiala, Ferozepur.
1982	Whitebacked plant hopper	PR107, PR 4141	Patiala, Ferozepur Kapurthala.
	Thrips	HM95, PR103	Kapurthala, Gurdaspur
	Yellow stem borer	PR4141, PR106	Ferozepur .
1983	Brown plant hopper, whitebacked plant hopper	PR 196, PR4141, Pusa-150, Pusa-169	Patiala, Ferozepur Kapurthala.
	Yellow stem borer	PR4141, PR 106, Basmati 370	Ferozepur
		Punjab Basmati 1	Kapurthala
	Thrips	PR106, Jaya,IR8	Ludhiana, Kapurthala
		PR106, PR414, Punjab Basmati 1	All rice growing areas.
	Hispa	PR 103,PR 106, PR 4141,IR8,Jaya, Basmati 370, Punjab Basmati 1.	Kapurthala, Gurdaspur.

Source : Sidhu

In Punjab, the experience with the new varieties was no better. They created new varieties of pests and diseases. The Taichung Native I variety which was the first dwarf variety introduced in 1966, was susceptible to bacterial blight and white backed plant hopper. In 1968 it was replaced by IR-8 which was considered to be resistant to stem rot and brown spot, but proved to be susceptible to both these diseases. Later varieties such as PR 103, PR 106, PR 108, PR 109 which were released after the failure of earlier dwarf varieties were specially bred for disease and insect resistance. PR 106, which currently accounts for 80% of the area under rice cultivation in Punjab, was considered resistant to white-backed plant hopper and stem rot disease when it was introduced in 1976. It has since then become susceptible to both, as well as to the rice leaf folder, hispa, stem-borer and several other insect pests.[19]

The 'miracle' varieties displaced the diversity of traditionally grown crops and through the erosion of diversity, the new seeds became a mechanism for introducing and fostering pests. Indigenous varieties, or land races are resistant to locally occurring pests and diseases. Even if certain diseases occur, some of the strains may be susceptible, while others will have the resistance to survive. Crop notations also help in pest control. Since many pests are specific to particular plants, planting crops in different seasons and different years causes large reductions in pest populations. On the other hand, planting the same crop over large areas year after year encourages pest build ups. Cropping systems based on diversity thus have a built-in protection.

As Howard noted half a century ago,

'Nature has never found it necessary to design the equivalent of the spraying machine and the poison spray for the control of insect and fungus pests. It is true that all kinds of diseases are to be found here and there among the plants and animals of the forest, but these never assume large proportions. The principle followed is that plants and animals can very well protect themselves even when such

things as parasites are to be found in their midst. Nature's rule in these matters is to live and let live.'[20]

Howard believed that the cultivators of the East had a lot to teach the Western experts about disease and pest control and to get Western reductionism out of the vicious and violent circle of 'discovering more and more new pests and devising more and more poison sprays to destroy them'. When Howard came to Pusa in 1905 as the Imperial Economic Botanist to the Government of India, he found that crops grown by cultivators in the neighbourhood of Pusa were free of pests and needed no insecticides and fungicides.

'I decided that I could not do better then watch the operations of these peasants and acquire their traditional knowledge as rapidly as possible. For the time being, therefore, I regarded them as my professors of agriculture. Another group of instructors were obviously the insects and fungi themselves. The methods of the cultivators, if followed, would result in crops practically free from disease, the insects and fungi would be useful for pointing out unsuitable varieties and methods of farming inappropriate to the locality.'[21]

At the end of five years of tuition under his new 'professors' – the peasants and the pests – Howard had learnt :

'... how to grow healthy crops, practically free from disease, without the slightest help from mycologists, entomologists, bacteriologists, agricultural chemists, statisticians, clearing-houses of information, artificial manures, spraying machines, insecticides, fungicides, germicides, and all the other expensive paraphernalia of the modern experiment station.'[22]

Howard could teach the world sustainable farming because he had the humility to learn it first from practicing peasants and Nature herself. The Green Revolution experts, in contrast, felt they can control and conquer nature through their 'miracle' seeds and chemicals. The new seeds and chemicals destabilize the farm ecology and create pest outbreaks in a number of ways. Firstly, the shift from organic to chemical fertilizers reduces the plants resistance to pest attacks. Thus there is a linkage between heavy use of fertilizers and vulnerability to pests. The excessive fertilizer uptake of the new varieties is found to contribute to disease vulnerability. Sidhu[23] reports that increase in N level from 0 to 12 kg/ha has been found to increase infestation by rice hispa from 68.7 to 171.1 damaged leaves/10 hills. Similarly, infestation by leaf folder increased from 13.9 to 43.3 damaged leaves/10 hill with an increase of nitrogen from 0 to 150 kg N/ha. Even those high yielding varieties of crops, which are specially bred for disease resistance become highly susceptible to certain types of diseases when heavy doses of fertilizers are applied. Such diseases, which usually attack the young succulent tissues, include the rusts, muts, powdery and downy mildews and the virus.

Chemical fertilizers, which are an essential part of the package of the new seed technology thus contribute to pest vulnerability by reducing resistance. The reduction in the genetic base from which the new varieties are developed also contributes to pest vulnerability, even when pest-resistance is part of the plant breeding strategy. Varieties may not be permanently resistant, because pests can change. A limited or even shrinking gene pool faces pests which continue to adapt through mutation, a process often increased by pesticide use. Pest resistance is an ecological state, not an engineered one. As Chaboussou comments, 'The gene, the vector of heredity, can only operate as a function of the environment. Thus it is useless improving the resistance of a plant to such and such a disease if that "genetic" immunity is going to be impaired by applying a pesticide aimed at some other pest.'[24]

While it is possible to engineer resistance in a plant at the laboratory level, the tendency for its breakdown in the field is a problem. The CGIAR's 1979 Integrative Report accepted that 'breakdown in resistance can occur rapidly and in some instances replacement varieties may be required about every three years or so.' While Howard found that pests are not a problem in ecologically balanced agriculture, in an unstable agricultural system, they pose a serious challenge to agronomy. The metaphor for pesticide use in agriculture then becomes war, as an introduction to a textbook on pest-management illustrates :

> 'The war against pests is a continuing one that man must fight to ensure his survival. Pests (in particular insects) are our major competitors on earth and for the hundreds of thousands of years of our existence they have kept our numbers low and, on occasions, have threatened extinction. Throughout the ages man has lived at a bare subsistence level because of the onslaught of pests and the diseases they carry. It is only in comparatively recent times that this picture has begun to alter as, in certain parts of the world, we have gradually gained the upper hand over pests.

> 'The war story described some of the battles that have been fought and the continuing guerilla warfare, the type of enemies we are facing and some of their manoeuvres for survival; the weapons we have at our command ranging from the rather crude ones of the "bow and arrow" age of pest control to the sophisticated weapons of the present day, including a look into the future of some "secret weapons" that are in the trial stages, the gains that have been made ; and some of the devastation which is a concomitant of war.'[25]

But the 'war' with pests is unnecessary. The most effective pest control mechanism is built into the ecology of crops, partly by ensuring balanced pest-predator relationships through crop diversity and partly by building up resistance in plants. Organic manuring is now being shown to be critical to such a building up of resistance.

The Green Revolution strategy fails to see the ecology of pests as well as that of pesticides because it is based on subtle balances within the plant and invisible relationships of the plant to its environment. It therefore simplistically reduces the management of pests to the violent use of poisons. It also fails to recognize that pests have natural enemies with the unique property of regulating pest populations.

In de Bach's view,

'The philosophy of pest control by chemicals has been to achieve the highest kill possible, and per cent mortality has been the main yardstick in the early screening of new chemicals in the lab. Such an objective, the highest kill possible, combined with ignorance of or disregard for, nontarget insects and mites is guaranteed to be the quickest road to upset resurgences and the development of resistance to pesticides.'[26]

De Bach's research on DDT-induced pest increase showed that these increases could be anywhere from thirty-six fold to over twelve hundred-fold. The aggravation of the problem is directly related to the violence unleashed on the natural enemies of pests. Reductionist science which fails to perceive the natural balance, also fails to anticipate and predict what will happen when that balance is disturbed. For example, insects and pests which were considered insignificant in Punjab before the Green Revolution have now become major problems. At present, rice cultivation in Punjab is vulnerable to about 40 insects and 12 diseases. The rice leaf folder, enaphalorrocis medinalis, was first recorded as a minor infestation in 1964. However, in 1967 it had appeared in epidemic form in Kapurthala, and has since then appeared in all rice growing areas of the state and caused heavy losses in 1983. The whitebacked plant hopper, Sogatella fureifers, was first recorded in 1966. Since then, severe outbreaks of the insect have occurred in 1972, 1975, 1978, 1981, 1982 and 1983.The brown plant hopper, Nilanparvate lugene, was first observed in 1973. In 1975, it appeared in an epidemic form in Kapurthala, Gurdaspur and Ferosepur. Rice thrips or stenchartothrips biformis, has become a major pest since 1982, when it appeared in Kapurthala and Gurdaspur. The earcutting caterpillar, Mythimn separate, has appeared all over Punjab during 1983. The ceecal life beetle, the horned caterpillar, the yellow hairy caterpillar, small brown plant hopper, and sugarcane pyrilla are the new insect pests of rice created by the Green Revolution. Among the new diseases to which the rice monocultures are susceptible are brown spot, false smut, sheath rot.[27]

Having destroyed nature's mechanisms for controlling pests through the destruction of diversity, the 'miracle' seeds of the Green Revolution became mechanisms for breeding new pests and creating new diseases. The treadmill of breeding new varieties runs incessantly as ecologically vulnerable varieties create new pests which create the need for breeding yet newer varieties. The only miracle that seems to have been achieved by the Green Revolution is the creation of new pests and diseases, and with them the ever increasing demand for pesticides. Yet the new costs of new pests and poisonous pesticides were never counted as part of the 'miracle' of the new seeds that modern plant breeders had gifted the world in the name of increasing 'food security'.

Table 2.8: Insect pests and diseases recorded on rice crop in Punjab.

S.No.	Common name	Scientific name	First record in punjab (Year)	Current pest status
1	2	3	4	5

INSECTS

S.No.	Common name	Scientific name	First record in punjab (Year)	Current pest status
1.	Rice root weevil	*Echinocnemus oryses* Marshall	1953	++
2.	Rice leaf-folder	*Cnaphalocrocis medinalis* (Guenee)	1964	+++
3.	Paddy caseworm	*Nymophula depunctalis* Gurer	1964	+
4.	Pink stem borer	*Sesamia inferens* (walker)	1964	++
5.	Rice thrips	*Stenchaetothrios biformis* (Bagnall)	1964	++
6.	Whitebacked planthopper	*Soqatella furcifera* (Harvath)	1966	+++
7.	Green Leafhopper	*Nephotettix niqrooictus* (Stao)	1966	++
8.	Green Leafhopper	*Nephotettix virescenus* (Distant)	1966	++
9.	Big white leaf hopper	*Tettigella spectra* Distant	1966	+
10.		*Nisia atrovensa* Lath.	1966	+
11.		*Oleovus.sp*	1966	+
12.	Maize borer	*Chilo zonellus* Swin.	1966	+
13.	Rice thrips	*Haplothrios qanglbaueri* Schmutz	1967	+
14.	Leaf hopper	*Exitianus indicus* Distant	1969	+
15.	Leaf hopper	*Cicadulina bipunctella* Matsumura	1969	+
16.	Leaf hopper Walker	*Parabolocratus porrectus*	1969	+
17.	Leaf hopper	*Thaia subrufa* Melichar	1969	+
18.	Cearal leaf-flea beetle	*Chaetocnema basalis* 1970	1972	+
19.	Horned caterpillar	*Melanitis leda ismene* (Gamer)	1971	+
20.	White hairy catepillar	*Euproctis virquncula* Walker	1971	+
21.	Yellow hairy catepillar	*Psalis pennatula* Fabricius	1971	+
22.	Brown Planthopper	*Nilaparvata lugens* (Staol)	1973	+
23.	Small brown planthopper	*Laodelphax striatellus* (Fallen)	1978	+
24.	Whorl maggot	*Hydrellia* spp.	1980	+
25.	Hydrophilid beetle	*Laccobius* spp.	1981	+
26.	Rice hispa	*Dicladispa armiqera* (oliver)	-	+++

1	2	3	4	5
27.	Stem borer white	*Scirpophaga innotate* (walker)	-	+++
28.	Yellow stem borer	*Scirpophaga incertulas* (walker)	-	+++
29.	Ear-cutting caterpillor	*Mythimna separate* (walker)	-	++
30.	Rice grasshopper	*Hieroglyphus banian* (Fabririus)	-	++
31.	Surface grasshopper	*Oxya nitidula* (Walker)	1981	++
32.	Rice bug	*Leptocrisa acuta*	-	++
33.	Toka	*Chroteqonus* app	-	+
34.	Rice Skipper	*Parnara mathias* Fabricius	-	+
35.	Big white leafhopper	*Kola mimica* Distant	-	+
36.	Zig-zag leafhopper	*Recilia dorsalis* (Motsch.)	-	+
37.	Blue beetle	*Altica cvanea* Weber	-	+
38.	Sugarcane pyrilla	*Pyrilla perpusilla* Walker	-	+
39.	Termite	*Microtermes obesi* (Holmgren)	-	+
40.	Termite	*Odontotermes obesus* (Rambur)	-	+

DISEASES

1	2	3	4	5
41.	Sheath blight	*Rhizoctonia solani* Kuhn	1960	+++
42.	Bacterial blight	*Xanthomonas campestris* p.v. *oryzae* (Ishiyama 1922) Dye 1978	1965	+++
43.	False smut	*Ustilaginoidea virens* (Cke.) Tak.	1975	++
44.	Sheath rot	*Sarocladium oryzae* Sawada	1978	+++
45.	Stem rot	*Sclerotium oryzae* Cav.	-	+++
46.	Brown spot	*Helminthosporium oryzae* van Breda dehan	-	++
47.	Blast	*Pyricularia oryzae* Cav.	-	+
48.	Karnel smut(bunt)	*Tilletia barclayana* (Bref.)	-	+
49.	Leaf smut	*Entyloma oryzae* H&P. Sydow	-	+
50.	Narrow brown leaf	*Cercospora oryzae* Miyake spot	-	+
51.	Bacterial leaf streak	*Xanthomonos translucens* f. sp. oryzae Pordesimo	-	+
52.	Leaf scald	*Rhynchosporium oryzae* Has & Yokogi	-	+
53.	Foot rot	*Gibberella fujikuroi* (saw.) Ito	-	+

+++ = Major
++ = Minor
+   = Insignificant

Source : *Sidhu*

References 2

1. Jack Doyle, *op cit*, p258.

2. Jack Doyle, *op cit*, p256.

3. Erna Bennett, 'Threats to Crop Plant Genbtic Resourses', in J G Howkes, *Conservation and Agriculture*, London: Duckworth, 1978, p114.

4. World Bank, National Seeds Project III.

5. Mahabal Ram, *High Yielding Varieties of Crops*, Delhi: Oxford, 1980, p212.

6. Lappe and Collins, *ibid*.

7. A K Yegna Iyengar, *Field Crops of India*, Bangalore: BAPPCO, 1944 (reprinted 1980), p30.

8. M S Swaminathan, *op cit*, p113.

9. C H Shah, (ed), *Agricultural Development of India*, Delhi: Orient Longman, 1979, pxxx ii.

10. R H Richaria, paper presented at CAP Seminar on 'Crisis in Modern Science', Penang, Nov 1986.

11. Yegna Iyengar, *op cit*, p30.

12. Geertz cited in T.B. Bayliss Smith and Sudhir Wanmali, 'The Green Revolution at micro scale', *Understanding Green Revolutions*, Cambridge University Press, 1984.

13. Dan Morgan, *op cit*, p36.

14. A H Church, *Food Grains of India*, Delhi: Taj Offset (reprinted), 1983.

15. D S Kang, 'Environmental Problems of the Green Revolution with a focus on Punjab, India' in Ricard Barrett, (ed), *International Dimensions of the Environmental Crisis*, Boulder: WestviewPress, 1982, p198.

16. Mahabal Ram, *op cit*.

17. M S Gill, Success in the Indian Punjab, in J G Hawkes, *Conservation and Agriculture*, London: Duckworth, 1978, p193.

18.   Bharat Dogra, *Empty Stomachs and Packed Godowns*, New Delhi, 1984.

19.   G S Sidhu, 'Green Revolution in Rice and its Ecological Impact-Example of High Yielding Rice Varieties in the Punjab', mimeo, p29.

20.   Howard, *op cit.*

21.   Howard, *op cit.*

22.   Howard, *op cit.*

23.   Sidhu, *op cit.*

24.   F Chaboussou, 'How Pesticides Increase Pests,' *Ecologist*, Vol. 16, No. 1, 1986, pp29-36.

25.   W W Fletcher, *The Pest War*, Oxford: Basil Blackwell, 1974, p1.

26.   De Bach, *Biological Control by Natural Enemies*, London: Cambridge University Press, 1974.

27.   Sidhu, *op cit*, p20.

3

CHEMICAL FERTILIZERS AND SOIL FERTILITY

PUNJAB has rich alluvial soils, characteristic of much of the Indo-Gangetic plains of North India.

Of these soils, Howard and Wad had said:

'...field records of ten centuries prove that the land produces fair crops year after year without falling in fertility. A perfect balance has been reached between the manurial requirements of the crops harvested and the natural processes which recuperate fertility.'[1]

And in his presidential address to the Agriculture section of the Indian Science Congress, G Clarke had said,

'When we examine the facts, we must put the Northern Indian cultivator down as the most economical farmer in the world as far as the utilization of the potent element of fertility, nitrogen, goes. He does more with a little nitrogen than any farmer I ever heard of. **We need not concern ourselves with soil deterioration in these provinces. The present standard of fertility can be maintained indefinitely.'**[2]

Twenty years of Green Revolution agriculture, have succeeded in destroying the fertility of Punjab soils which had been maintained over generations for centuries and could have been indefinitely maintained if international experts and their Indian followers had not mistakenly believed that their technologies could substitute land, and chemicals could replace the organic fertility of soils. It has been the assumption of the Green Revolution that nutrient loss and nutrient deficit can be made up by the use of non-renewable inputs of phosphorous potash and nitrates as chemical fertilizers. The nutrient cycle, in which nutrients are produced by the soil through plants, and returned to the soil as organic matter is thus replaced by linear non-renewable flows of phosphorous and potash derived from geological deposits, and nitrogen derived from petroleum.

Voracious Varieties

The Green Revolution was essentially a seed-fertilizer package since the new seeds were bred to be high 'consumers' of fertilizer. In the years prior to the Green Revolution, there was an excess of fertilizer capacity in the industrialised countries. After World War I, manufacturers of explosives, whose factories were equipped for the fixation of nitrogen, had to find other markets for their products. Synthetic fertilizers provided a convenient 'conversion' for peaceful uses of war products. Howard identified this conversion as closely linked to the 'NPK mentality' of chemical farming.

'The feature of the manuring of the west is the use of artificial manures. The factories engaged during the Great War in the fixation of atmospheric nitrogen for the manufacture of explosives had to find other markets, the use of nitrogenous fertilizers in agriculture increased, until today the majority of farmers and market gardeners base their

manurial programme on the cheapest forms of nitrogen (N), phosphorous (P), and potassium(K) on the market. What may be conveniently described as the NPK mentality dominates farming alike, in the experimental stations and in the countryside. Vested interests entrenched in time of national emergency, have gained a stranglehold.' [3]

After the Wars, there was cheap and abundant fertilizer in the west, and American companies were anxious to ensure higher fertilizer consumption overseas to recoup their investment. The fertilizer push was an important factor in the spread of the new seeds, because wherever the new seeds went, they opened up new markets for chemical fertilizers, In 1967, at a meeting in New Delhi, Borlaug was emphatic about the role of fertilizers in the new revolution. 'If I were a member of your parliament', he told the politicians and diplomats in the audience, 'I would leap from my seat every fifteen minutes and yell at the top of my voice, "Fertilizers!..., Give the farmers more fertilizers". There is no more vital message in India than this. Fertilizers will give India more food.' [4]

Government policy, inspired by international agencies had actively supported the use of chemical fertilizers through the Green Revolution. Chemical fertilizers had been subsidized, even given away by international agencies. The Ford Foundations' intensive agricultural district programme in India which started in 1952 was largely based on the intensive use of fertilizer. The World Bank and US AID were also involved in fertilizer pushing. In the 1960s these agencies began applying pressure on India to encourage western chemical companies to build fertilizer plants. Despite these investments in a fertilizer industry, India was dependent on imports for as much as 40% of its fertilizer needs created by the Green Revolution (Figure 7). [5]

Figure 7 : Production, Imports and Consumption of
Fertilizers, 1952-53 to 1975-76

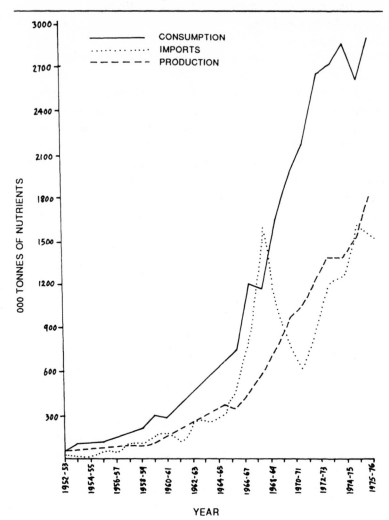

Source : Desai, 1979

The higher consumption of chemical fertilizers was a planned strategy in Indian agriculture policy. While the First plan viewed fertilizers as supplementary to organic manures, the Second and subsequent Five Year plans gave a direct and crucial role to fertilizers. The High Yielding Varieties programme depended crucially on higher inputs of fertilizer.

The NPK mentality had finally been transferred to Indian experts and farmers. Chemical fertilizers were viewed as being superior substitutes for the organic fertility of soils.

As one expert stated:

'Whenever a community experiences relative scarcity of a factor, it tries to evolve a substitute for it. Substitution of man by machine and of land by irrigation or fertilizers are common examples.... Fertilizers, the by-product of oil, similarly helped to substitute land which proved a constraint....' [6]

The conclusion was that 'the seed-fertilizer revolution is land augmenting'. There were two processes by which such land saving was assumed to take place. Firstly, the dwarf varieties were bred in such a way that they could consume three or four times higher doses of chemical fertilizer to produce higher quantities of grain than traditional varieties. Traditional crop varieties, characterised by tall and thin straw typically convert the heavy doses of fertilizers into overall growth of the plant, rather than increasing grain yield. Commonly, the excessive growth of the plant causes the stalk to break, 'lodging' the grain on to the ground, which results in heavy crop losses. The main characteristics of the 'miracle seeds' or 'high yielding varieties' which started the process of the Green Revolution was that they were biologically engineered to be dwarf varieties to avoid

Table 3.1:   Production and Imports of Fertilizers,
1952-53 to 1975-76
('000 tonnes of nutrients)

Year	Production	Imports	Production +Imports	Columns (3) as percent of col. (4)
(1)	(2)	(3)	(4)	(5)
1952-53	60	47	107	44
1953-54	67	26	93	28
1954-55	82	31	113	27
1955-56	89	61	150	41
1956-57	98	71	168	42
1957-58	107	123	230	54
1958-59	112	120	232	52
1959-60	135	179	314	57
1960-61	166	197	363	54
1961-62	219	174	393	44
1962-63	282	282	564	50
1963-64	327	274	601	46
1964-65	474	325	699	46
1965-66	357	492	849	58
1966-67	455	847	1,302	65
1967-68	610	1,623	2,233	73
1968-69	776	1,078	1,854	58
1969-70	955	762	1,717	44
1970-71	1,061	633	1,694	37
1971-72	1,239	971	2,210	44
1972-73	1,385	1,218	2,603	47
1973-74	1,374	1,256	2,630	48
1974-75	1,517	1,608	3,125	51
1975-76	1,828	1,541	3,369	46

lodging under conditions of intensive inputs of chemical fertilizers.

The second property for which these Green Revolution varieties were bred was their photo insensitivity. Photosynthesis – the combining of carbon dioxide in the air with water by the action of sunlight to form carbohydrates – is a normal and essential part of the process of plant growth. The uptake of very large doses of nitrates requires increased solar radiation, so that designing the plant for maximisation of leaf surface exposed to sunlight becomes a necessary complement of high-level fertilizer application. A high tillering rate and erect leaves are the structural changes to increase photosynthesis. Non-photosensitivity has also been bred in to the dwarf varieties so that they mature within a given period, unaffected by annual variation in the climate or by local variation in day-length, and multiple cropping becomes possible. Higher fertilizer application and multiple cropping both demanded intensive water use and perennial irrigation. Together with high water use, high fertilizer use and multicropping were seen as affecting a land-saving quality, and providing mechanisms for overcoming land scarcity. However, the technology which was aimed at increasing soil fertility and 'augmenting' land had the impact of reducing the availability of fertile and productive land.

The dramatic increase in cereal production in the Green Revolution state of Punjab has been linked exclusively to the use of this seed-fertilizer-package. However, the increase involved a number of factors including an increase in the cropped area, a shift from mixtures of cereals and pulses to monocultures of wheat and rice, a change from crop-rotations to multicropping of rice and wheat, and finally, the change in grain-straw ratio of the new varieties.

The erosion and degradation of land is intimately linked with cropping patterns of the Green Revolution. As a direct consequence of the Green Revolution's introduction in the Punjab, a very rapid change in the pattern of land-use has taken place. Crops are being grown in newly opened areas which were formerly forests or marginal lands. Croplands are now kept constantly under soil depleting crops like wheat and rice, rather than being rotated with soil building leguminous crops like pulses. As Kang has cautioned, 'This process implies a downward spiraling of agricultural land use – from legume to wheat to rice to wasteland.'[7]

The area under wheat has nearly doubled and the area under rice has gone up by five times since the start of the Green Revolution. During the same period, the area under pulses (legumes) has been reduced by one-half.

In 1982-83, Punjab had 2,552,248 acres under Jowar and 3,218,248 under Bajra. These minor cereals occupied 41 percent of the food grain area in Punjab. During that period, the area under wheat was 6,734,357 acres, and under rice was 775,367 acres. 85% of the area is under irrigation. 84% of the rabi cropped area is under wheat and 51% of the Kharif cropped area is under rice. This extensive crop cultivation of only two crops, with a very narrow genetic base, and with heavy inputs of irrigation and chemicals (fertilizers, insecticides, pesticides) is creating serious ecological problems in Punjab.

The changes in the cropping pattern in Punjab has resulted in a break in the recycling of nutrients in two ways. Firstly, the engineering of dwarf varieties leads to a reduction of the organic matter available for recycling into the soil, either directly or through the fodder cycle. Secondly, it demands higher nutrient uptake causing, simultaneously, the build up of toxic chemicals in the soil and micronutrient deficiencies.

The decline in acr·age under pulses and coarse grains and the increase in area under hybrid wheat and rice has a serious impact on the fertility of soil. The removal of pulses from the cropping pattern removes a major source of free nitrogen for the soil. Decline in millets and course-grain production leads to decline in fodder and hence farm yard manure which is an important source for the renewal of soil fertility.

For equivalent fertilization, the high yielding varieties produce about the same total biomass as traditional rice. They increase the grain yield at the cost of the straw. Thus, while traditional rice produces four to five times as much straw as grain, high yielding rice typically produce a one to one ratio of grain to straw. Thus a conversion from traditional to high yielding rice increases the grain available but decreases the straw. The scarcity in straw ultimately reduces biomass availability for fodder and mulch, leading to a breakdown in nutrient recycling.

Traditional varieties of sorghum yield 60 maunds of straw per acre while producing 10 maunds of grain. On, the other hand modern rice varieties produce equivalent amounts of grain and straw, thus reducing fodder and organic fertilizer supply.

The principle which had protected the fertility of the alluvial soils of the Indo-Gangetic plains over centuries was the ecological principle of balancing growth with decay at the organic level, based on treating the soil as a living system.

As Howard has indicated:

'Agriculture must always be balanced. If we speed up growth we must accelerate decay. If, on the other hand, the soil's reserves are squandered, crop production ceases to be good farming, it becomes something very different. The farmer is transformed into a bandit.' [8]

Twenty years of 'Farmers' Training and Education Schemes' have taught the Punjab farmer how to be efficient bandits. 64% of all crop loans in Punjab have gone to financing the purchase of fertilizer in the state. Markfed, the state co-operative marketing federation distributes almost Rs55 million worth of fertilizers every year. Fertilizer consumption in the state has increased thirty fold since the inception of the Green Revolution.

Table 3.2 to 3.4 indicate the increase in fertilizer application involved in a shift to HYV as prescribed by the Punjab Agricultural University.

Has this increased fertilizer use been reflected in increased production? Before the introduction of dwarf wheats, the yield of the local high yielding varieties was 3291 Kgs/ha, as against 4690 Kgs/ha of the dwarf varieties. The recommended N,P,K for the two varieties was 60:40:20 and 120:60:60 respectively. The physical output-input ratio has thus actually declined from 55 to 40 for N, from 82 to 78 for P and from 165 to 78 for K with a shift to the new seeds. The increased use of chemical fertilizers for new seeds has not therefore corresponded to an equivalent increase in output.The shift from organic to chemical fertilizers and the associated shift from traditional high yielding varieties has lowered the inherent productivity of soils by creating new deficiencies and diseases. As a result, there is a stagnation in the response of crops to chemical fertilizer application. The reductionist Green Revolution paradigm which saw chemical fertilizers as substitutes for land and the inherent fertility of soils, failed to perceive that plants need more than NPK for growth. What was viewed as 'underutilisation' of land by the Ford Foundation's IADP was, in effect, the use of land within sustainable limits of the renewability and recycling of all nutrients. And the 'intensive' agriculture strategy was, in effect, a robbery of the soil's fertility.

To use Pyarelal's words,

'So much soil fertility was bartered away for commercial gain, without the possibility of returning in any shape or form to the soil what was taken out of the soil, thus

Table 3.2 : Fertilizer recommendations for irrigated wheat over years.

Period	Variety	Nutrient (Kg/ha)		
		N	P_2O_5	K_2O
Uptill 1968-69	Tall & semi dwarf	45	23	23
since 1969-70	i)Tall Varieties (Local)	60-80	40	20
	ii)Dwarf Varieties (HYV)	120	60	30

Table 3.3 : Fertilizer recommendation for rice over the years

Period	Variety	Nutrient (Kg/ha)		
		N	P_2O_5	K_2O
Uptill 1968-69	Tall Varieties	45	23	23
since 1969-70	Dwarf Varieties	120	60	60
	Tall Varieties	60	30	—
since 1975-76	Dwarf Varieties	120	30	30
	Tall Varieties	60	20	—

Table 3.4: Fertilizers recommendations for irrigated maize over the years

Period	Variety	Nutrients (Kg/ha)		
		N	P_2O_5	K_2O
Uptill 1968-69	Hybrids	110	55	55
	Locals	60	25	30
since 1969-70	Hybrids	120	60	60
	Locals	85	60	30
since 1975-76	Hybrids	120	60	30
	Locals	85	30	20

*impairing it permanently. This is not agriculture but
down right robbery of the soil at the cost of posterity.'* [9]

Diseased and Dying Soils

After a few years of bumper harvests in Punjab, crop
failures at a large number of sites were reported, despite
liberal applications of NPK fertilizers. The new threat came
from micronutrient deficiencies caused by the rapid and
continuous removal of micronutrients by 'high yielding'
varieties. Plants quite evidently need more than NPK, and
the voracious high yielding varieties draw out micronutrients from soils at a very rapid rate creating micronutrient
deficiencies of zinc, iron, copper, manganese, magnesium,
molybdenum, boron, etc. With organic manuring these
deficiencies do not occur because organic matter contains
these trace elements, whereas chemical NPK does not. Zinc
deficiency is the most widespread of all micronutrient deficiencies in Punjab. Over half of the 8706 soil samples from
Punjab exhibited zinc deficiency which has reduced yields of
rice, wheat and maize by up to 3.9 tonnes, 1.98 tonnes, 3.4
tonnes per ha respectively. Consumption of zinc sulphate
rose from zero in 1969-70 to nearly 15,000 tonnes in 1984-
85.[10] Iron deficiency has been reported in Punjab, Haryana,
Andra Pradesh, Bihar, Gujarat, and Tamil Nadu, and is
hurting yields of rice, wheat, ground nut, sugarcane, etc.
Manganese is another micronutrient which has become
deficient in Punjab soils. Sulphur deficiency which was
earlier noticed only in oilseed and pulse crops has now been
noticed in cereals like wheat (Figure 8).

The Green Revolution has also resulted in soil toxicity by
introducing excess quantities of trace elements in the ecosystems. Fluorine toxicity from irrigation was introduced in
various regions of India. 26 Mha of India's lands are affected

Figure 8 : Micronutrient deficiencies in
Punjab soils

 Zn [single line represents 20% area deficient in a block]

 Fe [single 'x' represents 10% area deficient in a block]

 Mn [visual symptoms and response]

 Probable areas of Fe, Mn and S deficiency

Source : *Punjab Agricultural University*

by aluminium toxicity. In Hoshiarpur district of Punjab, boron, iron, molybdenuym and selenium toxicity has built up with Green Revolution practices and is posing a threat to crop production as well as animal health.

As a result of soil diseases and deficiencies, the increase in NPK application has not shown a corresponding increase in output in rice and wheat. The productivity of wheat and rice has been fluctuating and even declining in most districts in Punjab, inspite of increasing levels of fertilizer application.

Experiments at the Punjab Agricultural University (PAU) are now beginning to show that chemical fertilizers cannot be substitutes for the organic fertility of the soil, and organic fertility can be maintained only by returning to the soil part of the organic matter that it produces. In the early1950s, before the entry of the advisors of the Ford Foundation, when KM Munshi referred to repairing the nutrient cycle, he was anticipating what agricultural scientists are today recommending for the diseased and dying fields of Punjab. And Howard's predictions are beginning to come true that, 'In the years to come, chemical manures will be considered as one of the greatest follies of the industrial epoch.'

The Return to Organic Inputs

Agricultural scientists are calling for a return to organic inputs to maintain crop productivity.[11] What the Ford Foundation and other experts saw as 'shackles of the past' are once again being recognised as timeless, essential elements of sustainable agriculture.

Green manuring, an ancient practice in rice cultivation, has been found to double the response to nitrogenous fertilizers. Green manuring combined with 60 kg N/ha, produced biomass of rice seedlings equal to that produced with 120 kg N/ha. In regions such as Karnataka, pongamia leaves and branches are applied generously to paddy fields for improving soil fertility and soil structure.

Similarly, the age-old practice of applying farmyard manure has been shown by the PAU to be more effective than chemical fertilizers. Farmyard manure (FYM) applied at the rate of 12 t/ha to rice in rice-wheat system increased the rice yield by 0.8 t/ha and when applied with 40 and 80 kg N/ha the increase in yield was 1.8 and 2.9 t/ha respectively as compared to 2.76 t/ha yield with 120 kg N/ha alone.

The yield of wheat after rice with 12 ton FYM added to 90 kg N and 30 kg P_2O_5/ha was comparable to that obtained with the recommended rates of 120 kg N and 60 kg P_2O_5/ha.

Results of long term experiment revealed that application of 15 t FYM/ha to maize in maize-wheat rotation resulted in a saving of 40, 80, 60, 30 and 5 kg of N, P_2O_5, K O and Zn respectively. Also, the experiments at the cultivator's fields in different parts of the state revealed a substitution of 50% NPK. Continuous application of FYM has been shown

Table 3.5 : Farm Yard Manure (FYM) and fertilizer economy in rice-wheat cropping system

FYM (t/ha) to rice	Rice		Wheat		
	N applied (Kg/ha)	Yield (t/ha)	N Kg/ha	P_2O_5 Kg/ha	Yield (t/ha)
0	0	3.9	0	0	1.4
12	0	3.7	. 90	0	3.0
12	40	4.7	90	30	3.4
12	80	5.8	90	30	4.3
0	120	5.6	120	60	4.2

to improve soil productivity as a result of increase in soil organic matter content. It has also been found to remove micronutrient deficiency by making available P, K and Zn.

Sustainable agriculture is based on the recycling of soil nutrients. This involves returning to the soil, part of the nutrients that come from the soil either directly as organic fertilizer, or indirectly through the manure from farm animals. Maintenance of the nutrient cycle, and through it the fertility of the soil, is based on this inviolable law of return.

The Green Revolution paradigm substituted the nutrient cycle with linear flows of purchased inputs of chemical fertilizers from factories and focused on the production of marketable agricultural commodities. Yet, as the Punjab experience has shown, the fertility of soils cannot be reduced to NPK from factories, and agricultural productivity necessarily includes returning to the soil part of the biological products that the soil yields. Technologies cannot substitute nature and work outside nature's ecological processes without destroying the very basis of production. Nor can markets provide the only measure of 'output' and 'yields'.

The Green Revolution created the perception that soil fertility is produced in chemical factories, and agricultural yields are measured only through marketed commodities. Nitrogen fixing crops like pulses were displaced. Millets which have high yields from the perspective of returning organic matter to the soil, were rejected as 'marginal' crops. Biological products not sold on the market but used as internal inputs for maintaining soil fertility were totally ignored in the cost-benefit equations of the Green Revolution miracle. They did not appear in the list of inputs because they were not purchased, and they did not appear as outputs because they were not sold.

Yet what is 'unproductive' and 'waste' in the commercial context of the Green Revolution is now emerging as productive in the ecological context and as the only route to sustainable agriculture. By treating essential organic inputs that maintain the integrity of nature as 'waste', the Green Revolution strategy ensured that fertile and productive soils are actually laid waste. The 'land-augmenting' technology has proved to be a land-degrading and land-destroying technology. With the problem of greenhouse effect and global warming, a new dimension has been added to the ecologically destructive effect of chemical fertilizers. Nitrogen based fertilizers release nitrous oxide to the atmosphere which is one of the Greenhouse gases causing global warming. Chemical fertilizers have thus contributed to the erosion of food security through the pollution of the land, water and the atmosphere.

References 3

1.   Alfred Howard in M K Ghandi, *Food Shortage and Agriculture*, Ahmedabad: Navjivan Publishing House, 1949, p183.

2.   C G Clarke, in M K Ghandi, *ibid*, p83.

3.   Alfred Howard, *Agricultural Testament*, London: Oxford, 1940, p 25.

4.   Jack Doyle, *Altered Harvest*, p259.

5.   Gunvant Desai, 'Fertilizers in India's Agricultural Development', C H Shah, *Agricultural Development of India*, Orient Longman, 1979, p390.

6.   C H Shah, *op cit*, pxxxiii.

7.   D S Kang, 'Environmental Problems of the Green Revolution with a focus on Punjab, India', in Richard Barrett (ed), *International Dimensions of the Environmental Crisis*, Boulder: WestviewPress, 1982, p204.

8.   Howard, *op cit*, p26.

9.   Pyarelal in M K Ghandi, *Food Shortage and Agriculture*, p185.

10.   Punjab Agricultural University, Department of Soils, mimeo, 1985.

11.   Punjab Agricultural University, Department of Soils, mimeo, 1985.

4

INTENSIVE IRRIGATION, LARGE DAMS AND WATER CONFLICTS

Thirsty Seeds

WHEREVER the 'miracle' seeds of the Green Revolution went, they created a new thirst for water. Intensive chemicals and intensive irrigation were the two means used in Green Revolution agriculture to 'augment' land and improve soil fertility. Instead, they created land degradation and hence land scarcity, even while they created an addiction to pesticides, fertilizers and intensive water use. This chapter traces how intensive agriculture demanded intensive water use and created wasteland instead of increasing land productivity, and how it created new demands and unresolvable conflicts over water resources.

Punjab literally means the land of five rivers. The prosperity of the region is linked intimately with the sustainable use of the waters of the Indus (Sindhu) and its tributaries – the Jhelum (Vitasta), the Chenab (Asikin), the Ravi (Irravati), the Beas (Vapasa) and the Sutlej (Satadru).

Irrigation did not come to Punjab with the Green Revolution. Punjab has an ancient irrigation history. During the time of the Greek invasion, a flourishing agriculture existed in the region, served by a network of inundation canals. As far back as the 8th century, Arab conquerors differentiated between irrigated and non-irrigated lands for the purpose of levying taxes. Inundation canals irrigated millions of hectares. A great advantage associated with them was that they did not cause water logging. They flowed for only four to five months during the monsoon season, and for the remaining part of the year, were dry and served as drainage channels.

A second characteristic of the old canals was that they were aligned along natural drainage features. Every river which overflows its banks during the monsoon season, deposits silt on its banks. The recurring silt deposits build up a ridge high along the river. The run-off-the-river canals were built along these ridges. This alignment interfered very little with the general run-off of the rainfall in the region. The Sutlej Valley project, comprising of 13 large canals taking off from four large headworks, also followed the principle of alignment along natural drainage patterns of the land, not against them. In the 19th century, many old canals were provided with permanent headworks and made perennial. By the middle of the twentieth century, Punjab had as many as 31 large canal systems. To these was added the Bhakra dam in 1963.[1]

The Bhakra dam was conceived in 1908 with a 395 ft high reservoir level. In 1927 this was revised to 1600 ft. After independence, the Bhakra dam assumed a new significance because large areas under irrigation in the Indus Basin had gone to Pakistan. By 1953, the design of Bhakra dam was completed, and in 1963 the dam was ready, at a cost of Rs2,385 million.

Figure 9 : Indus Basin (India)

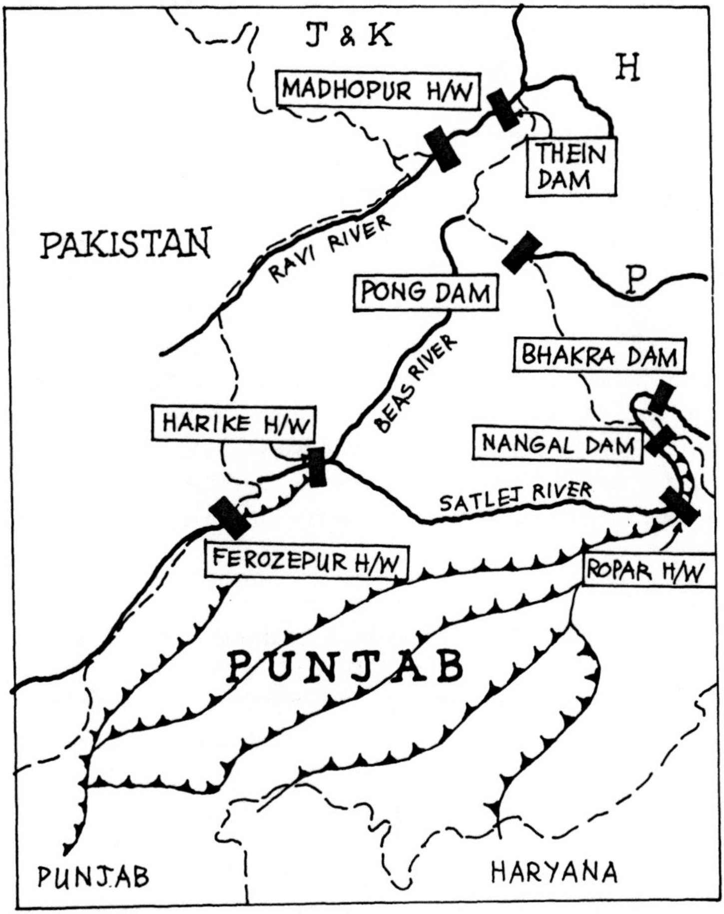

Source : *Uppal, 1978*

The Bhakra system covering a length of over 3000km of channels was different from older canal systems because it was fed by a high dam, and included a network of cross country canals which ran against the natural drainage to irrigate 2,372,100 hectares.

In 1977, the large dam network to feed the Indus irrigation system was expanded with the construction of the Pandoh dam near Mandi and the Pong dam near Talwara. The Pandoh dam has been built to divert 7000 cusecs of Beas water over 40km into the Sutlej River upstream of the Gobind Sagar lake of Bhakra dam. The Pong dam is meant to store 6.55 million acre feet of water for feeding the Rajasthan Canal. Another dam under construction is the 5.6 million acre feet capacity high dam at Thein on the River Ravi (Figure 9). This is meant to stabilise the flow of the Madhopur – Beas link, besides irrigating 24691 ha in J & K. The storage capacity of various dams in Punjab is given in Table 4.1

Table 4.1: Storage Capacity of dams constructed in Punjab

Dam	Gross	Storage M.Cu.M. Live
Bhakra (Govind Sagar)	9867.8	7770
Beas dam at Pong	9140.94	6907.49
Thein dam	7404.00	—

The combination of large dams for surface irrigation and the high water demands for Green Revolution agriculture has led to major ecological impacts on the water balance and political impact on the power balance of the region.

The Upper Jhelum and the Upper Chenab are principally feeder canals transferring the Jhelum water to the Chenab and the Chenab water across the Ravi. The Lower Chenab was opened as an inundation canal in 1887 and as a perennial canal in 1893. The Sirhind Canal was opened for irrigation in 1882 but was not completed till 1884.

The Green Revolution was based on the expansion and intensification of irrigation from surface as well as ground water. The new seeds have an enormous thirst for irrigation water. Compared to the earlier varieties needing protective irrigation as an insurance against crop failure, the new seeds need intensive irrigation as an essential input for crop yields. The Green Revolution increased the need for irrigation water at two levels. Firstly, the shift from water prudent crops such as millets and oilseeds to monocultures and multicropping such as wheat and rice increased the demand for water inputs throughout the year. Table 4.3 gives the water requirements of different crops.[2]

Table 4.3 : Productivity of cereals per unit of water

Crop (New strains)	Water Requirement in a typical tract (mm)	Yield Kg/ha	Water Use Effiency per mm water
Rice	1200	4500	3.7
Sorghum	500	4500	9.0
Bajra	500	4000	8.0
Maize	625	5000	8.0
Wheat	400	5000	12.5

Source: GIRIAPPA, p 17

Secondly, the replacement of old varieties of wheat with new varieties of wheat and rice also increased the intensity of irrigation, which went up from 20 - 30% to 200 - 300%.

Irrigation, which used to be protective, now became

Table 4.2 : Various perennial canal systems in the Punjab together with area under canal and irrigation

Canal	Date of opening	capacity as first designed (cs.)	sec.	maximum capacity 1921-22 (old canals)	command area C.C.A	area irrigated during the year 1921-22 (old canals)(ha)	length in distributary (miles)(ha)
1.	2.	3.	4.	5.	6.	7.	8.
1. Bikaner canal	1927	2720	77	20	2,83,200	225855	135
2. Eastern Canal	1938	3200	91	—	1,46,650	165032	—
3. Western Yamuna Canal	1939-40	6500	184	256/313			
4. Ferozepur Feeder	1952	1952	11100	315	315	22,865	12520
5. Sirhid Feeder		1952	4760	135	315	3,48,982	47162
6. Makhu Canal					3,24,396	9904	
7. Uppwe Bari Doab Canal(Remodelled)	1954	6700	190	595	384856		
8. Sirhind Canal (Remodelled)	1954			354			
9. Bist Doab Canal	1954	1601	45	45	2,02,201	91441	64
10.Nangal Hydel Channel	1954	12500	354	354	—	30373	1140

1.	2.	3.	4.	5.	6.	7.	8.
11. Bhakra Canals	1954	18000	510	including Bist	4 Million Doab Sirhind 27,11,400	2.6 Million	3360
12. Rajasthan Canal	1954	18500	524	524		590400	8000
13. Western Jamuna Canal	1973	2800	79	184	9,40,674	346163	3032
14. Upper Bari Doab	1859 & 1873	5000	142	190	6,08,684	543695	2510
15. Sirhind Canal	1882-84	6000	170	242	1,84,104	499240	5478
16. Sidhnai Canal	1886-87	1820	52	52	11,21,540	84747	402
17. Lower Chenab	1887 & 1893	8313	236	308	10,46,320	1036395	3589
18. Lower Jhelum Canal	1901-02	3800	108	119	5,06,637	359280	1674
19. Upper Chenab	1912	11742	333	138 338	6,11,360	265245	2003
20. LOwer Bari Doab Canal	1915	6750	191	200	5,77,600	402055	1923
21. Upper Jhelum Canal	1915	8380	237	57	2,32,023	145110	106
22. Lower Swa/Canal	1917-18			237			

'productive'. Green Revolution varieties need much more water than indigenous varieties. High Yielding Varieties of wheat, for example, need about three times as much irrigation as traditional varieties. Thus, while indigenous wheat varieties need 12 inches of irrigation, the HYV's require at-least 36 inches. The comparative yields of native wheat varieties and the HYV varieties is 3,291 and 4,690 Kg/ha respectively in Punjab.[3] The productivity with respect to water use is therefore 620.90 and 293.1 kg/ha/cm respectively.

From the perspective of water use, the shift to the new wheat varieties, and the replacement of millets and maize by rice has therefore led to a decrease in productivity. In addition, the shift has induced processes of social and ecological disruption. Social considerations of equity favour the extensive use of irrigation water which assures a protective dose of water to crops over as large an area as possible. The intensive use of irrigation as part of the Green Revolution packet limits the provisioning of irrigation to a smaller region. Thus a shift from millets to paddy amounts to a restriction of irrigation from 3 ha to 1 ha.

The intensive use of water also has major ecological impacts. The dramatic increase in water use with the Green Revolution has led to a total destabilisation of the water balance in the region. The water cycle can be destabilised by adding more water to an ecosystem than the natural drainage potential of that system. This leads to desertification through waterlogging and salinisation of the land. Desertification of this kind is a form of water abuse rather than water use. It is associated with large irrigation projects and water intensive cultivation patterns. About 25% of the irrigated land in the US suffers from salinisation and waterlogging. In India 10 million hectares of canal irrigated land have become waterlogged and another 25 million hectares are threatened with salinity. Land gets waterlogged when the water table is within 1.5 to 2.1 meters below the ground surface. The water table goes up if water is added to a basin faster than it can drain out. Certain types of soils and certain types of topography are most vulnerable to waterlogging. The rich alluvial plains of Punjab which have a very negligible slope suffer seriously from desertification induced by the introduction of excessive irrigation water to make Green Revolution farming possible.

The rise or fall in the ground water table in any region depends upon the water balance of that area. If the amount of water being added through deep percolation and seepage from irrigation systems is more than the water that drains out of the region either through lateral flow or tubewells, the water balance will be positive and the water table will continue to rise in that area. On the other hand, if withdrawal of water from the ground is more than the deep percolation losses, seepage and lateral movement to that area, the water balance of that region will be negative and the

Table 4.4: Rainfall in various districts of Punjab
(in mm)

District	1972	1973	1974	1975	1976	1977	1978
Gurdaspur	716	1057	600	954	1371	1211	903
Amritsar	442	773	379	543	1233	702	543
Kapurthala	410	690	339	620	593	610	*
Jullundur	514	881	368	649	627	775	646
Hoshiarpur	683	867	511	788	992	950	713
Ropar	552	837	657	806	629	732	816
Ludhiana	469	694	365	633	692	878	652
Ferozepur	301	532	172	350	764	445	360
Faridkot	319	543	237	411	608	514	426
Bhatinda	310	428	240	602	422	354	387
Sangrur	396	560	275	485	626	680	*
Patiala	571	728	419	660	882	944	650

* Not Known.

water table will continue to fall in that area climatically.
Punjab is semi-arid to arid. Table 4.4 gives the rainfall in
various district of Punjab.[4] The average rainfall from 1957-
81 observed at the Faridkot rain-gauge was 400mm. Most of
the rainfall normally occurs during the months of July to September. In the Indo-Gangetic plains, 20% of the rainfall
infiltrates the ground water body. Thus the vertical flow
from surface water to ground water under natural conditions is only 80mm annually. About 25-30% of the water
applied to field crops in the form of irrigation or rain
percolates downward and is added to the groundwater
table resulting in its rise. In the absence of a lateral flow of
sub-surface water, the deep percolation losses from fields
under different crop rotation and the rise of water table is
clear from the following sample calculations made by the
Punjab Agricultural University.[5]

Let the total water applied annually to fields with a
cotton-wheat or maize-wheat rotation be 130cm. It has been
observed that inspite of the judicious use of water, about 25
percent of the applied water percolates down and joins the
groundwater. This means that about 32cm depth of water
percolates downwards. Assuming an average specific yield
of 0.2 and in a confined aquifer, it will result in a 160cm rise
in the water table every year. The specific yield is the quantity of water that can be drained by gravity per unit volume.
Such an annual downward loss of water with a rice-wheat
rotation goes as high as 50% of the applied water. Annual
water applied to a rice-wheat crop rotation is 200cm and
thus 100cm will find its way into groundwater. In a confined
aquifer the annual rise in the groundwater table will be
500cm.

In addition to the contribution due to cropping pattern
changes, the other important cause for the rise of the water
table in the south-west part of Punjab is the sub-surface flow

of water towards that region. Groundwater aquifers in the Punjab State are not strictly isolated regionwise. There is a lateral flow of groundwater from a north-east to south-west direction. The flow gradient except in the Kandi tract is about 0.3 meter per kilometer. Uppal and Mangat, have shown that the depth to water table contours are advancing from north-west Punjab towards Bhatinda with an average rate of travel of about 0.29 km per year and are causing the water table to rise in this tract. A groundwater ridge has developed roughly along the Bikaner canal in the south-west part.

The canal system existing in the Punjab State is shown in Figure 10. The main canals that feed the south-west region originate from Ropar Head Works, Harike Barrage and Ferozepur Head Works. The water level in the area has increased steadily with the construction of the Bhakra System and Harike Barrage. There has, however, been a demand to increase the canal supplies because of increasing irrigation requirements as the intensity of cropping has increased faster than that envisaged when the canal supplies were allocated.

During the last few years, with the rise of the water table in this region, the area under rice cultivation has increased many fold. Intensive irrigation thus starts a vicious cycle of demand for more water. Further, since the underground water quality in the Southwest of Punjab is saline, the farmers are demanding more and more canal water inspite of water-logging conditions. Several areas in the state of Punjab are affected by waterlogging and salinity problems. It is esti-mated that an area of about 2.86 lakh hectares has a water table depth of less than 1.5 meters even in the month of June. The water table further rises by 0.5 to 1.2 meters during the monsoon season. These areas are normally subjected to waterlogging problems of varying degrees depending upon the topography of the area. Table 4.5 gives the district wise groundwater balance in Punjab.[6] The water table depth in

Figure 10 : Surface irrigation system of Punjab

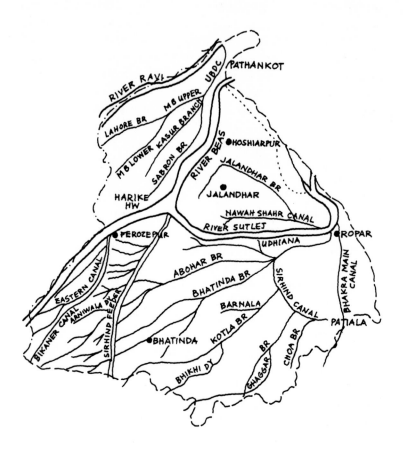

Source : *Punjab Agricultural University*

Figure 11 : Distribution of waterlogged areas in Punjab

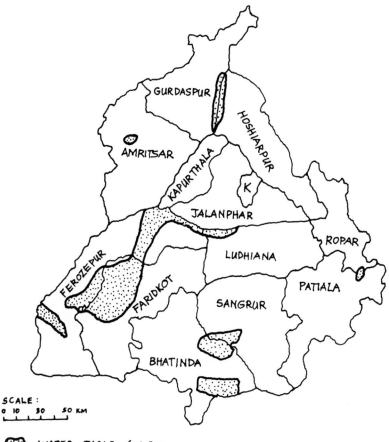

SCALE :
0   10   30   50 KM

WATER TABLE < 1·5 m
IN JUNE 1983

Source : *Punjab Agricultural University*

Figure 12 : The quality of underground irrigation waters in Punjab

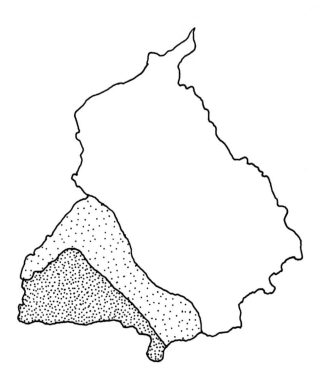

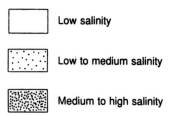

☐ Low salinity

▨ Low to medium salinity

▨ Medium to high salinity

Source: *Punjab Agriculture University*

Table 4.5: Districtwise groundwater balance (hectare metres) in Punjab

District	Recharge due to rainfall	Recharge due to canal irrigation and seewage from canal system	Recharge due to tubewell irrigation	Recharge due to Drain to river sub soil flow from high water table areas	Draft due to Tube wells	Dug wells	Water balance
AMRITSAR	107541	74947	22105	473	88102	850	116614
GURDASPUR	126485	50182	14849	14550	59006	391	146668
HOSHIARPUR	41380	15861	12634	7661	48232	1809	27495
KAPURTHALA	28597	1817	12964	11208	51703	154	2729
JULLUNDUR	48543	19067	26452	18869	101137	4679	702
LUDHIANA	24999	45825	22151	5847	85445	1263	12112
SANGRUR	39085	56039	22697	—	88870	1921	27030
ROPAR	20657	11365	5321	3041	19088	1952	19345
PATIALA	76313	48144	30786	5796	121920	1691	37429
FARIDKOT	43449	97173	13349	—	53306	92	100573
BHATINDA	26731	63688	8567	—	34243	44	64699
FEROZEPUR	53860	92747	20318	17567	30905	981	102606
PUNJAB	367640	576855	212193	85120	831957	15828	658002

different regions of Punjab is shown in Figure 11 and the
distribution of waterlogged areas in different districts of
Punjab (water table less than 1.5 meter in the month of June,
1983)is given in Table 4.6.[7] It may be seen from the map and
Table 4.6 that the major problem of water logging is in the
South-west districts of Punjab i.e. Faridkot, Ferozepur and
Bhatinda. Faridkot and Ferozepur districts alone have about
2.14 lakh hectares areas which has water table less than 1.5
meter depth and is affected by salinity and sodicity problems. Bhatinda is the next most affected district in Punjab.

Table 4.6 : Distribution of waterlogged areas in different districts of Punjab
(Water table less than 1.5m June, 1983)

S.No.	District	Waterlogged area lac hectares	percentage in each
1	Faridkot	1.12	39.16
2.	Ferozepur	1.02	35.66
3.	Bhatinda	0.32	11.19
4.	Sangrur	0.09	3.15
5.	Amritsar	0.08	2.80
6.	Hoshiarpur	0.07	2.45
7.	Gurdaspur	0.06	2.10
8.	Jalandhar	0.05	1.75
9.	Ludhiana	0.04	1.40
10.	Ropar	0.005	0.17
11.	Patiala	0.005	0.17
	Total	2.86	

Closely related to the problem of waterlogging is the
salinity creation. The salt poisoning of arable land, seems
to be an inevitable consequence of intensive irrigation in
arid regions. In regions of scarce rainfall, the earth contains
a large amount of unleached salts. Pouring irrigation water
into such soils brings those salts to the surface and leaves
behind a residue when the water evaporates. Today more

than one third of the world's irrigated land has salt-pollution problems that diminish the productivity of the soil and, in extreme cases, ruin it forever. It is estimated that about 0.7 lakh hectares in Punjab are salt affected and produce either no yields or very poor yields (Figure 12).

Waterlogging and salinity are problems linked to the overuse of water in regions where the nature of the topography and soils rule out intensive irrigation as a productive use of land and water. The natural ecological solution to these problems would be to shift to more water prudent cropping patterns, to crops and varieties that need less water. The engineering solution, on the other hand is to redesign nature by artificially transforming the drainage characteristics and chemical composition of soils. The cure is worse than the disease; more water consumption, more drains to get rid of excess water quickly, more energy and capital for desalting. These cures are neither affordable nor sustainable.

The demand for more water by Punjab is linked to the demands to treat waterlogging and salinity by using more water rather than by decreasing the high inputs of water to supply the intensive need of the Green Revolution crops. All responses to waterlogging are attempts to treat the symptom, not the cause. It is assumed that intensive water use must continue and engineering solutions must be found for removing excess water.

First, the solution offered was 'vertical drainage' i.e. pumping out ground water by tubewells. In this case, the solution to ecological problems of irrigation is more expensive than the original costs of irrigation. The cost of irrigation is less in the case of canal water as compared to tubewell water. The cost also varies with diesel and electric operated tubewells. The cost per irrigation for maize-wheat or cotton-wheat rotations in the case of a tubewell operated by diesel is about 3 times higher as compared to an electric-operated tubewell. Because of this differential cost, even in the areas where water table is at the depth of 1 metre, farmers are demanding more and more canal water. The cost of installing tubewells using 'skimming well technology' to treat waterlogging is much higher than the cost of irrigation. This technology is used so as not to disturb the deep brackish water and to prevent the mixing of fresh water with deep saline water. The installation cost increases with the increase in the number of wells and the spacing between the wells primarily because of the cost of connecting pipes, digging trenches and boring wells.

Table 4.7 : Cost calculations for PVC material (6 kg/cm2) installation for single well and multiple well systems

Single/multiple wells (4)	Approximate cost Rs. to single well	Relative increase in cost compared
Single well lower upto 22m depth. 4-multiple well system	2,000	—
i) 6m spacing	3,800	1.90
ii) 12m spacing	4,700	2.35
iii) 18m spacing	5,600	2.80
iv) 24m spacing	6,500	3.25

Some tentative calculations of the costs involved in the installation of PVC-made tubewells for a single well of 22m deep, and for the 4 skimming wells system, each well of 6m deep at variable spacing between the wells, are given in Table 4.7.[8] It is seen that even for PVC material (pressure 6 KG/CM2) using the skimming well technique with 4-wells with a 6m spacing in between, the installation cost increase

1.9 times as compared to the single well technique. Increasing the spacing between the wells from 6m to 24m increases the cost of installation by 3.25 folds. Increasing the number of wells from 4 to 6 will further increase the installation cost. The vertical drainage method has been found not to be adopted by farmers due to these economic factors and now the horizontal drainage method is being experimented with. The average per hectare cost of installing the drainage system is in the range of Rs10,000 to Rs12,000 which is again way beyond the reach of most farmers. A Rs10 crore scheme, to reclaim 5,000 hectares of waterlogged area is now being taken up in Punjab on an experimental basis with assistance from the World Bank. The project involves the construction of a sub-surface horizontal pipe drainage system in 41 low lying areas scattered in the south-western districts of Faridkot, Ferozepur, Bhatinda and Sangrur.

Intensive irrigation also introduces conflicts between private and social interests. Waterlogging does not recognize farm boundaries, and drainage cannot be managed except as a community activity. But community management of resources has been the first casualty in the privatisation thrust of the Green Revolution. While lab scale solutions exist for the waterlogging problems, socio-ecologically they are unrealistic given their demands for financial inputs and social organisation. Ecological conflicts over water use in Punjab are thus not resolvable immediately.

It is estimated that out of the total geographical area of 50.38 lakh hectares of Punjab, 42 lakh hectares are under agriculture. 82% of the cultivated area is irrigated and the remaining 18% is rainfed. 58% of the irrigated area is irrigated by tubewells and 42% by canals. By increasing the intensity and scale of irrigation, the Green Revolution was to have been a strategy for transcending the limits on productivity set by natural conditions. However, instead of increasing the productivity of land, intensive water use has turned large parts of the region into a water logged desert. Where irrigation is dependent on ground water, the water table is declining at an estimated rate of one to one and a half foot every year, due to over-exploitation. This virtually amounts to mining of underground water resources of the state. Table 4.8[9] indicates the increase of tubewell irrigation in Punjab and Figure 13[10] shows the rapid growth of tube-well irrigation the state:

A recent survey by the Directorate of Water Resources Punjab has shown that 60 out of 118 development blocks in the state cannot sustain any further increase in the number

Figure 13: Punjab: Tube wells & open wells

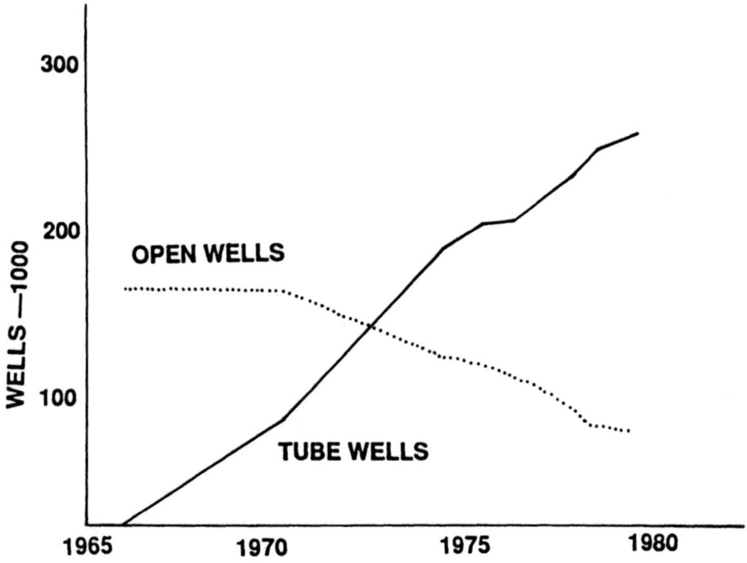

Source: *Statistical Abstract of Punjab (1961-79)*

Table 4.8 : Increase in tubewells in the Punjab

Year	Tubewells (Lakhs)		
	Diesel operated	Electric operated	Total
1970	1.01	0.91	1.92
1975	3.04	1.46	4.50
1976	3.03	1.67	4.70
1977	3.04	1.96	5.00
1978	3.02	2.33	5.35
1979	3.23	2.62	5.85
1980	3.20	2.80	6.00
1981	3.10	3.00	6.10
1982	2.90	3.33	6.23

Source : Sidhu

of tubewells. 34 other development blocks have only marginal possibilities of exploiting the ground water.

Energized pumping of ground water was central to the controlled input needs of Green Revolution seeds. The 'efficiency' was calculated only in terms of energy and horse power, not in terms of whether the withdrawal was in consonance with recharge to ensure the sustainable use of water. It is estimated that an irrigation pump powered by a 7.5 kg electric motor takes five hours and one man to irrigate an acre of wheat compared to 60 bullock hours and 60 man hours with a persian wheel. The calculation that was never carried out by Green Revolution experts was that the persian wheel has supported agriculture for centuries while the energized pump threatens to dessicate large areas of prime farmland in less than two decades.

The Green Revolution strategy for generating abundance by improving the productivity of land and water has thus turned into a strategy for creating scarcity of land and water and generating new conflicts. Both in terms of nutrients and water, the Green Revolution strategists held the view that they were 'freeing' themselves of nature's processes, central to which are the nutrient and water cycle. They assumed that:

> 'In the early stages of economic development when the supply of the "produced factor", capital, is modest and the level of technology is rudimentary a country's natural resource base looms large in determining the volume of agricultural output and the possibilities for high rates of savings. Advances in technology, capital accumulation, and international trade that come with economic growth progressively free the production process from a one-to-one relationship between output and specific resource inputs.'[11]

Yet the Punjab experience brings home the point that even the Green Revolution was bounded by ecological limits, and by attempting to break out of them, it further increased those limits, generating new levels of scarcity, insecurity and vulnerability.

Large Dams and the Centralisation of Political Power

Intensive irrigation systems create the need for large scale storage systems, and large dams create the need for the centralised control of water resources. The increased demand for water created by the new seeds has, on the one hand, created a chain reaction of ecological impacts, and on the other hand it has induced a chain reaction of water conflicts in the region. One of the most significant impacts of large dams in India is the destabilisation of catchment areas, which leads to the aggravated erosion and siltation of the reservoir. According to the 1972 irrigation commission report, the assumed annual rate of siltation for the Bhakra dam was 23,000 acre ft, while the observed rate was 33,745 acre ft. The assumed silt index was 105 acre ft per 100 sq. miles of catchment area, while the observed silt index was 154.

With increasing demands for water, and decreasing storage capacity, the waters of the Bhakra dam had to be augmented by waters from the Pong dam on the Beas. In December 1977, the Beas-Sutlej link project was commissioned to augment the flow to the Bhakra dam by diverting 7500 cusecs from the Beas to the Sutlej. The Beas-Sutlej link hydro-electric project in Himachal Pradesh has caused vast environmental changes in the region. Ever since the project was completed in 1978, the micro-climate has changed dramatically in Kulu, Mandi and Bilaspur districts.

A number of studies conducted by a team of scientists from Himachal Pradesh Agricultural University, Palampur, reveal that the rainfall has also decreased by 100mm to 200mm annually. A project meant to 'augment' water resources has thus actually contributed to an overall decline by the reduction in rainfall in the catchment. Major environmental changes have occurred due to the diversion of nearly 4,716 million cubic metres of water of the Beas from Pandoh in the Mandi district to Slapper in the Bilaspur district, a stretch of about 40km. From Pandoh the water flows through a 13.1km long tunnel to an open channel extending to 11.8 km and finally through another 12.38km long tunnel from Sundernagar to Slapper where it joins the Sutlej.

The extensive use of explosives to construct the tunnel has blocked the natural springs and waterways, causing a shortage of drinking water in the catchment of the Pandoh dam. Water scarcity has also been aggravated by the indiscriminate felling of trees, the blasting of rocks for the construction of the dam and the diversion. Thus, inspite of the Beas project, the Bhakra dam has remained vulnerable to erratic flows. In the summer of 1985, the dam had reached its dead storage level of 1462 ft. The inflows at Bhakra were around 10,000 cusecs, compared to 24,000 cusecs in 1984. The water level was falling at over one foot per day. The Beas-Sutlej link meant to bring 7500 cusecs to the Bhakra dam was only carrying 4,000 cusecs.[12] The vulnerability of the large dam system also increased due to sub-surface seepage. Satellite imagery has shown that there is a seepage zone of about 3km wide and 40km in length of water leaking into faults from the Pong dam.[13] This seepage threatens the safety of the dam.

Further, the centralised control of the Bhakra system had made the Indus basin more vulnerable to floods as well as to water scarcity, which have further fueled water conflicts between neighboring states, and between the states and the Centre.

In May 1984, the Bhakra Main Canal was breached near Ropar. Haryana saw it as an act of sabotage, and the Governor was asked to ensure the protection of the entire 250 kilometer long length of the canal in the Punjab territory. Haryana's share in the 15,000 cusec capacity canal is 9,500 and the breach had created a serious water crisis in the state. The breached Bhakra main-line canal is the life-line for Sirsa, Jind and Fatehabad districts of Haryana. The canal breach had forced the government to provide emergency supplies of drinking water by tankers. The cost to Haryana in terms of crop losses was estimated at Rs200 crores.[14]

In September 1988, Punjab was flooded (Figure 14). 65% of its 12,000 villages were marooned, and 34 lakh people in 10 of the state's districts were affected. The state suffered a loss of about Rs1,000 crores. 80% of the standing crop was destroyed, and 1,500 people were reported killed. As experts of PAU have shown, these deaths, and the floods, 'were very

Figure 14

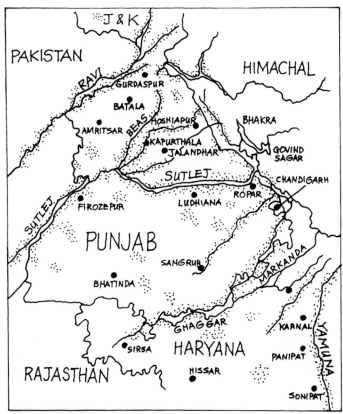

Source : *Times of India, Sept. 29, 1989*

Flood affected areas.

much man-made with a major share of the blame due to
BBMB'. The BBMB (Bhakra Beas Management Board) authorities had filled up the Bhakra dam up to the highest ever
water level of 1687.47 feet, which was 2.5 feet above the
maximum storage capacity, as early as 12 September, largely
for the Prime Minister's visit for the Bhakra Silver Jubilee
day. When the waters rose further, they released 3.8 lakh
cusecs of water into the Sutlej river, which was already

carrying 2 lakh cusecs against its capacity of 3 lakhs. Water was similarly released without warning from the Pong dam. The PAU experts have contended that the 'deluge in these areas was not entirely due to rains as was being made out but due to criminal water management by the BBMB who went about irrationally releasing water discharges in lakhs of cusecs without any warning to the thousands of people who live close to embankments of the two rivers.'[15]

Large dams like the Bhakra have been built to stabilise water flows and to free farmers from the 'vagaries' of the rainfall. Yet, the Punjab floods like those of 1988 remind us that dams, too, depend on rainfall, and can actually destabilize water systems instead of controlling water.

For the entire month of September, meteorologists had predicted 121mm of rainfall. But in just four days, between September 23 and 26, Bhakra received 85mm of rain in just 45 minutes, close to what the meteorological department had predicted for the entire month. The BBMB released water from the dams to save the dam, even through it destroyed human, animal and plant life. Between 25 and 28 September, the water released was 4 lakh cusecs, which was twice the total discharge of 2.5 lakh cusecs during all of August. Similarly, the water discharge from the Pong dam during the same period was over 7 lakh cusecs compared to only 1.5 lakh cusecs in the previous month. Ten feet high cascades of water from the two reservoirs washed away entire villages within hours. Roads disappeared under a sea of water. And the people had little idea of what was to come because the dams were controlled not by the people, not by the state government, but by the BBMB which is controlled by the Central government. This centralization has increased the ecological vulnerability of the river system.

Besides the ecological instability associated with large

dams for intensive water use for intensive agriculture, water intensive agriculture has also demanded the centralised control of water. Ecological vulnerability and centralised control of the irrigation system of Punjab has generated their own contribution to the violence in Punjab. On 7 November, 1988, the Chairman of the Bhakra Beas Management Board was shot dead by four people outside his residence in Chandigarh. The killing took place in the background of the Punjab floods for which the BBMB was seen as the agency responsible. Since BBMB is under the Central government, the floods aggravated the conflict between Punjab and the Centre.[16]

The older canal systems of Punjab were regionally managed within the State. A special circle of the PWD Irrigation Branch, known as Derajot circle was established in the Punjab in the 19th century to maintain the inundation canals. With the opening of the Bhakra system, a new centralisation in water control took place. This was formalised further with the setting up of the Bhakra Beas Management Board. When Punjab was bifurcated into Punjab and Haryana, through the Punjab Reorganisation Act, 1966, the management and control of river water shifted to the Centre.[17]

The Act empowers the Central government to set up two boards, namely; The Bhakra Management Board and the Beas Construction Board. The Bhakra Management Board was constituted by the Central government with effect from 1 October 1967. The Board is entrusted with the work of administration, maintenance and operation of the whole of Bhakra-Nangal irrigation and power complexes. The functions of the board include the regulation of the supply of water from Bhakra Nangal Project to the states of Haryana, Punjab and Rajasthan having regard to any agreement or arrangement between the governments of erstwhile State of Punjab and the State of Rajasthan and agreement between the sucessor states to the State of Punjab, namely, Haryana and Punjab. The functions include the construction of the remaining works of the Right Bank Power House.

The Board consists of a full-time Chairman and two full-time members appointed by the Central government; a government representative from the States of Punjab, Haryana and Rajasthan and the union territory of Himachal Pradesh, each nominated by the respective governments, and two representatives from the Central government.

The Act specified that the board shall be under the control of the Central government and it must comply with the directives given to it by the Central government. Furthermore, the Central government is empowered to issue directives to the relevant states in order to enable the Board to function effectively and the states are bound to comply with such directives coming from the centre.

As a result of the linguistic division of the State of Punjab, the construction of the Beas Project was taken over by the Central government on behalf of the successor states (Punjab and Haryana) and the State of Rajasthan. The funds for the construction were to be given by the concerned states. For discharging the function of constructing the project, the Punjab Reorganization Act, 1966, empowered the Central government could also issue mandatory directions to the concerned states in furtherance of the construction of the project. As and when the different phases of the project were completed, they were transferred to the Bhakra Management Board. So on the completion of the whole project, it would come under the management of the Bhakra Management Board which would be renamed as the Bhakra-Beas Management Board.

The idea of the Centre's control originated in the suggestion made by Prime Minister Nehru in a letter dated 3 July 1948, to the then Minister for Works, Mines and Power:

'The Bhakra Scheme is a big scheme and an urgent one, even more urgent than others. Thus far it has been carried on in a spasmodic way and what surprises me is that the Centre has little to do with it although we supply the entire finances. This is entirely unsatisfactory and I think we should make it clear that we cannot finance a scheme unless we have an effective voice in it. The East Punjab Government has to shoulder tremendous burdens and in the nature of things they cannot function as effectively as the Centre can.'[18]

Mega projects thus tend to centralize power and the loss of power by the federating units becomes a cause for conflict. The centralised control of the Punjab River system through the BBMB was reinforced by the changes introduced in the allocation of water by Indira Gandhi during the emergency in 1976, and again, after her return to power in 1981. In 1986, the Centre took further control over the water resources of Punjab by introducing the Inter-State Water Dispute (amendment) Bill 1986, to amend the Inter-State Water Disputes Act, 1956.

Irrigation continues to be a state subject under the Constitution of India which came into force in 1950. For the settlement of inter-state water disputes, the Draft Constitution of India initially contained identical provisions as the 1935 Act. But later, Parliament enacted the Inter-State Water Disputes Act, 1956. The Act provides for the constitution of a tribunal by the Central government for the settlement of an inter-state water dispute when a request is received from a state government, and when the Central government is of the opinion that the dispute cannot be settled by negotiations. Originally the Act provided for a one-man tribunal from among judges of the Supreme Court or a High Court, sitting or retired, nominated in this behalf by the Chief Justice of India. Later, the Act was amended to increase the membership of the tribunal to three sitting judges of the Supreme Court or High Court.

The 1986 Amendment to the Water Disputes Act was viewed as a further increase in power of the Centre over the states. However, under the earlier Act, one of the states had to refer the case to the Centre. Under the Amendment, the Centre can adjudicate in river water conflicts without one of the states referring the case to the Centre.

Inter-state Water Conflicts and the Elusive Search for Equity

Punjab, the land of five rivers, has been riddled with water conflicts for the past forty years. Politics and water distribution have been intimately intertwined in the region. On the one hand, the political fragmentation of Punjab has generated conflicts on the sharing of waters. On the other hand, the centralisation of water control has aggravated these conflicts. Another cause for the aggravation of these conflicts is the increased demand for water for Green Revolution agriculture. In this section we trace how the political divisions of Punjab created conflicts, and how the increasing centralisation of control over water distribution to provide inputs for the new seeds deepened these conflicts.

Punjab has been politically divided twice in less than half a century. It was first divided on the basis of religion in 1947 at the time of partition when West Punjab went to Pakistan and East Punjab came to India. It was divided again on a linguistic basis in 1966, when Punjab and Haryana were formed out of the erstwhile Punjab state. These divisions, combined with the rising demands for water intensive agriculture, have generated increasing conflicts over the Punjab rivers, between the three states of Punjab, Haryana, and Rajasthan, and between Punjab and the Centre.

At the time of the partition in 1947, arose a conflict over the sharing of the Indus waters. Out of the irrigated area of 9.6 mha of the pre-partition Indus basin, 8mha had gone to Pakistan and 1.6 mha came to India. Long drawn out negotiations between the governments of India and Pakistan, under the auspices of the World Bank, resulted in the Indus Waters Treaty in 1960.[19] Under this treaty, the waters of three Eastern rivers, Sutlej, Ravi and Beas of 3.1×10^{10}m^3/yr was allocated to India, in lieu of the surrender of India's claims to the water of the western rivers to Pakistan. The division of the Indus waters after full implementation of the Indus Waters Treaty will be 17.9×10^{10} m^3/yr, of the region's other three rivers – Indus, Jhelum and Chenab compared to 9.8×10^{10} m^3/yr of canal diversions in the 1960s. Prior to this, on the basis of an interstate conference on the 'Development and Utilisation of the waters of the Rivers Ravi and Beas' held in 1955, the Central government in India had allocated the surplus of 15.85 million acre feet (MAF) of these rivers, over and above the actual pre-partition utilisation as follows:[20]

East Punjab	:	7.20 MAF
Rajasthan	:	8.00 MAF
J & K	:	0.65 MAF

While the Indus Waters Treaty resolved the conflict over the sharing of Indus waters between India and Pakistan, the conflict between Punjab, Haryana and Rajasthan has continued to become increasingly intractable inspite of many agreements. The dispute between Punjab, Rajasthan and Haryana functions at many levels – there is conflict over the quantity of water available, over the just share of each state and the mode of distribution of this water.

In order to distribute and utilise the waters of the Ravi and Beas rivers between the three states, it was necessary to construct storage reservoirs and canals. A major project called the Beas project, a multipurpose scheme, was conceived as a joint venture of the states of Rajasthan consists of two units, the Beas dam at Pong and the Beas-Sutlej link fed by the Pandoh dam. The construction of the Pong dam was required at Beas mainly for storing waters for the 6.55 MAF of the Rajasthan canal which takes off from the Harike barrage. The Rajasthan Canal, 692km long, with a full supply discharge of 528 cusecs of water is the largest irrigation channel in the world. When completed it will irrigate eight million hectares of parched land of the Thar desert. The Beas-Sutlej link project is a power-cum-irrigation project and its purpose is to divert about 3.8 MAF of water from the Pandoh dam on the Beas River to the Sutlej River for an onward flow into the Bhakra lake. Part of this water is to be transported to Haryana through the Sutlej Yamuna Link Canal (SYL) (Figure 15).

After the partition of Punjab into Punjab and Haryana, a dispute arose between the two states over their respective shares in the water allocated to the composite state of Punjab. Haryana claimed 4.8 MAF out of the total of 7.2 MAF allocated to composite Punjab, and Punjab claimed the entire water. Punjab wanted the entire quantity of water on two grounds: (1) that the rivers flow through Punjab; (2) that the waters of the rivers were to be made available for irrigation through a canal system which lies entirely in the reorganised Punjab. The Punjab Government disallowed a survey of certain areas in Punjab by Haryana engineers. Haryana, on the other hand, has asserted that Punjab has no locus standi in the matter as the surplus waters of Ravi and

Figure 15

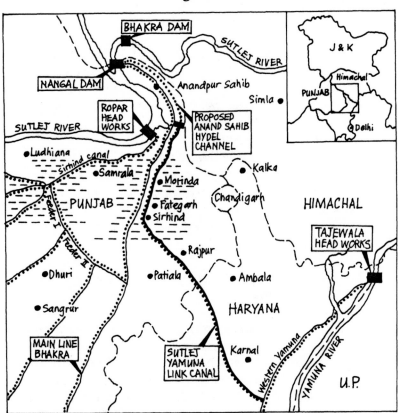

Source: *India Today, Nov 30 1985*

Beas have been acquired by the Central government on payment of compensation to Pakistan. Haryana also bases its claim on 'needs and principles of equity'. In the light of the failure of Punjab and Haryana to arrive at any agreement in the matter of sharing of the Ravi-Beas waters, the Haryana Government intimated the Government of India and requested the Central government to determine the shares in accordance with the provision of Section 78(1) of the Punjab

Reorganisation Act 1966, which provides for apportionment of the rights and liabilities of the Bhakra Nangal and Beas Projects. Section 78 of the Act reads as follows:-

> 'Notwithstanding anything contained in this act but subject to the provisions of Sections 79 and 80, all rights and liabilities of the existing State of Punjab in relation to Bhakra Nangal Project and Beas Project shall, on the appointed day, be the rights and liabilities of the successor states in such proportion as may be fixed, and subject to such adjustments as may be made, by agreement entered into by the said states after consultation with the Central Government, or if no such agreement is entered into within two years of the appointed day, as the Central Government may by order determine having regard to the purposes of the projects.'[21]

To resolve the conflict, there were fact-finding committees, Planning Commission interventions and also recommendations by the Chairman of the Central Water Commission. Finally, the Central government issued an order on 24 March, 1976 about the apportionment of surplus Ravi-Beas waters in accordance with the provisions of Section 78 of the Punjab Reorganisation Act, 1966, according to which 3.5 MAF of water was allocated to Haryana and the balance, not exceeding 3.5 MAF, to Punjab, out of the total surplus of Ravi-Beas water of 7.2 MAF falling to the share of the erstwhile state of Punjab after setting aside 0.2 MAF for the Delhi Drinking Water Supply.

This decision was taken during the emergency, and was resisted by the people and government of Punjab. Resistance to the decision was also expressed in the Anandpur Sahib Resolution, presented at the Ludhiana Conference of the Akali party of Punjab in 1978. Resolution No. 2 of the Anandpur Sahib Resolution states:

2(a) The control of head works should continue to be vested in Punjab, and if need be, the Reorganization Act should be amended.

2(b) The arbitrary and unjust Award given by Mrs Indira Gandhi during the Emergency on the distribution of Ravi-Beas waters should be revised on the universally accepted norms and principles thereby allowing justice to be done to Punjab.[22]

In 1977, the Congress was defeated in the general elections, largely as a consequence of the emergency. The then Punjab Government under the Akali Dal sought a review of the notification for a higher allocation of the Ravi-Beas waters. The Haryana government, meanwhile, filed a petition in the Supreme Court praying, inter-alia, that a directive be issued to Punjab for expeditiously undertaking the construction of the Punjab portion of the Sutlej-Yamuna Link (SYL) canal and declaring that the 1976 notification of the Centre allocating 3.5 MAF of Ravi-Beas waters each to Punjab and Haryana, was final and binding. Punjab also approached the Supreme Court questioning the constitutionality of Section 78 of the Punjab Reorganisation Act of 1966 as also the 1976 notification, apportioning the surplus Ravi-Beas flows between the two states. This case further delayed the resolution of the conflict.

After the break-up of the Janata Party, the Lok Sabha elections and Mrs Gandhi's return to power at the Centre, the Akali-Janata coalition in the Punjab which enjoyed majority support was removed, and President's rule was imposed. After a short spell, elections were held to the Punjab assembly and the Congress emerged victorious. With the states of Punjab, Rajasthan and Haryana all under Congress Rule, negotiations were opened and the suits filed by Punjab and Haryana were withdrawn by the new Congress

Chief Ministers under orders from Mrs Gandhi. The order of 1976 was replaced by the accord signed by the Chief Ministers of Haryana, Punjab and Rajasthan on 31 December, 1981. According to this agreement, the new surplus Ravi-Beas water was calculated as 17.17 MAF, based on the flow series from 1921-60 of the Beas Project Report compared to the corresponding figure of 15.85 MAF for the flow series for 1921-45 which had formed the basis of water allocation under the 1955 decision. The Chief Ministers of Punjab, Haryana and Rajasthan agreed on 31 December 1981, that the main supply of 17.17 MAF (flow and storage) be reallocated as follows,

Share of Punjab	:	4.22 MAF
Share of Haryana	:	3.50 MAF
Share of Rajasthan	:	8.60 MAF
Quantity earmarked for Delhi Water Supply	:	0.20 MAF
Share of J & K	:	0.65 MAF
TOTAL	:	17.17 MAF

The Chief Ministers further agreed that until such time as Rajasthan was in a position to utilise its full share through the Rajasthan Canal (now renamed the Indira Gandhi Canal), Punjab should be free to utilise the waters surplus to Rajasthan's requirements. Thus, during this period Punjab's share would be 4.82 MAF. Haryana's share was, however, left untouched.

Mr Bhajan Lal, the Haryana Chief Minister, protested against this, but relented when Mrs Gandhi incorporated Clause IV in the agreement to make it obligatory for Punjab to construct the 6500-cusec SYL Canal in its territory by 31 December, 1983 at the latest. It was further laid down that the alignment of the Punjab portion would be finalised by 31 March, 1981. Failing this the Centre was to undertake to do so on its own, so as to finalize the alignment by mid-April at the latest, with the Centre's decision binding on both Punjab and Haryana. At Mr Bhajan Lal's insistence, a condition was attached to the agreement that work on those portions of the SYL Canal whose alignment had already been finalised would commence within 15 days of the signing of the agreement. The monitoring of the construction of the SYL canal was to be undertaken by the Centre. Punjab raised various issues like the hardship caused to the families whose lands were to be acquired for the SYL. Haryana agreed to bear the cost of suitably rehabilitating these families on the stipulation that on no account would the construction of SYL be delayed by Punjab.

With the incorporation of all these conditions in the agreement, Mr Bhajan Lal signed it, and Mrs Indira Gandhi formally launched the construction of the SYL at Kapoori village in Punjab on 8 April, 1982. The very next day the Akalis in Punjab launched an agitation to prevent the construction of SYL Canal. Subsequently, the Akali leader Longowal announced a movement called the 'Nahar Roko' or 'Stop the Canal' movement, to prevent the digging of the Canal. In March 1986, the Damdami Taksal announced that it would launch 'Kar-Seva' to fill in that part of SYL which has already been dug. This followed the signing of the Rajiv-Longowal Accord in July 1985, in which the latter committed the Akali Dal to complete the SYL Canal by 15 August, 1986. Clause 9 of the Longowal Accord exclusively aimed at the sharing of River Waters, states:

'9.1. *The farmers of Punjab, Haryana and Rajasthan will continue to get water not less than what they are using from the Ravi-Beas system as on 1.7.1985. Waters used for consumptive purposes will also remain unaffected.*

Quantum of usage claimed shall be verified by the Tribunal referred to in para 9.2 below.

'9.2. The claims of Punjab and Haryana regarding the shares in their remaining waters will be referred for adjudication to a Tribunal to be presided over by a Supreme Court Judge. The decision of this Tribunal will be rendered within six months and would be binding on both parties. All legal and constitutional steps required in this respect be taken expeditiously.'

'9.3. The construction of the SYL canal shall continue. The canal shall be completed by 15th August 1986.' [23]

The Rajiv-Longowal Accord at resolving the river conflicts generated its own new conflicts. Both Rajasthan and Haryana felt it was a violation of their role as the real parties involved in the Interstate conflict. A tribunal headed by Justice V Balakrishna Eradi was set up by the Ravi and Beas Waters Tribunal Ordinance issued on 24 January, 1986. On 30 March, 1986, the ordinance was repealed and a new bill was passed called the Inter-State Water Disputes (Amendment) Bill 1986. This amendment to the Inter-State Water Disputes Act. 1956 was needed by the Centre because under the latter, the Central government could not adjudicate in river-water conflicts without one of the states referring the case to the Centre. For the states of Rajasthan and Haryana, the Punjab Accord and the amendment to the Inter-State Disputes Act was an erosion of the autonomy of the states, vis-a-vis the Centre.

Rajasthan refused to accept the portion of the Centre-Akali accord dealing with sharing of river waters. Opposition leaders described clause 9 of the accord, which deals with the river waters issue, as a 'direct assault' on the interests of Rajasthan. Opposition leaders warned that the 'fire had been extinguished at one place, Punjab, only to be lighted in Rajasthan'. They also protested against the amendment of the Interstate-Water disputes Act which gives new power to the Centre. In the Rajasthan Vidhan Sabha, the opposition members protested against the new Bill and called it a 'conspiracy by the Centre and the Punjab government'. Mr Nanak Chand Surana of the Janata Party said that the bringing of the new tribunal under the purview of the inter-state river water dispute act was fraught with dangerous consequences. Under the new Bill the Centre assumed all the powers to make whatever changes it liked in the future as regards the sharing of the river waters since the tribunal's judgement could not be challenged even in the Supreme Court. This, he said, was in sharp contrast to the Eradi Commission, whose decision could be challenged in the Supreme Court. The opposition alleged that making amendments in the original Act to deprive states of the right to have a say in the appointment of a tribunal was against the federal character of the country and made a 'mockery of the Constitution'.[24]

In Haryana, the Rajiv/Longowal Accord led to the creation of the Haryana Sangarsh Samiti 'to save and secure the interest of Haryana' which had been 'sold-out with the Union Government entering the Rajiv-Longowal accord'. Condemning the 'Rajiv-Longowal Accord' as a complete surrender by the Centre to the interests of Punjab and as totally anti-Haryana, the Sangarsh Samiti, under the leadership of Devi Lal, sought the abrogation of clause 9 of the accord dealing with distribution of river waters, at a massive rally called the 'Samast Haryana Sammelan' organised on 23 March,1986.[25]

In Punjab, the rebel Akali faction under Badal saw the Barnala government as siding with the Centre, while the Barnala government was itself unhappy with the working of the Eradi Tribunal set up to resolve the water conflict. When the Eradi Tribunal was to submit its verdict on the Ravi-Beas waters, the political parties in Punjab decided to launch a people's movement against it. The Akali Dal (Badal) decided to launch a 'people's movement' against any verdict by the Eradi Tribunal. In a lengthy resolution, the working committee said the terms of reference of the Eradi Tribunal had excluded the riparian principles. This discrimination had left Punjab with only 5.2 million acre feet (MAF) out of the 32 MAF of waters in the state's rivers. The resolution stated that 'We do not wish other states to be deprived of water, but they can get only what is surplus after Punjab's needs are met'. It added that with its present terms of reference, the decision of the tribunal would be 'irrelevant to Punjab's future needs'. It called for the scrapping of all water agreements made after 1955.[26]

The resolution also said the party had hoped the pressure exerted by it and the people of the state would force the Barnala Government to boycott the proceedings of the tribunal if the Centre did not incorporate the riparian rights of Punjab in its terms of reference. The water conflict in Punjab was thus rendered even more intractable. Punjab, Haryana and Rajasthan were all unhappy with the Accord, and with the Eradi Tribunal. Each state was now embroiled in a Centre-state conflict over the sharing of river waters even while the inter-state conflicts had deepened. In Punjab, there was every effort to block the construction of the SYL Canal, while in Haryana, politics centred on its construction.

The 213km long canal, meant to carry Haryana's share of the Ravi Beas waters to irrigate about 3 lakh hectares for Green Revolution agriculture is bogged down in a political quagmire. The cost of the Canal has shot up from 46 crores at the time of inauguration in 1976 to 456 crores in 1988. Over Rs362.58 crores has already been spent, and the canal is nowhere near completion.

In October 1986 irate farmers in the Ropar district of Punjab where the SYL canal takes off, virtually forced the Irrigation Department to abandon work on the project. And in May 1988, 30 Bihari labourers were killed at one of the project sites of the canal.

The farmers, who were sitting in dharna at Patheri Jattan village in Ropar, the nerve-centre of the agitation against the SYL canal were worried that the SYL canal would ruin them: either by taking away their fertile land or by causing water-logging. The present alignment of the canal runs parallel to the Bhakra main line canal from Lohand Khad near Anandpur Sahib to Beera Majri near Morinda, a distance of 67 kms. The land that the proposed canal will run through in Ropar, Morinda, Kurali, Kharar, Mohali, Chandigarh is the most fertile tract in Punjab and the canal would adversely hit 1,200 families. Of these, 300 families would be rendered homeless and the holdings of 700 others would become uneconomical. Speaker Ravi Inder Singh, whose constituency falls in this area, says: 'This will bring utter ruin to these families. Our earlier experience is that most farmers blow up the money they get as compensation. They are not attuned to any other vocation'.[27]

The problem is indeed real because many families had ended up as paupers when their land was acquired for the construction of the Bhakra main canal. Several displaced farmers had become drug addicts while many others had turned into alcoholics.

Karnail Singh, of Stahpur village, who is threatened with displacement reports :

'We had inherited 16 acres of which eight acres were ac-

quired for the Bhakra main canal. The Government is now
threatening to acquire the other eight acres also. We could
not use the previous money properly as no land was avail-
able. We have no plans to use this money also.'

The farmers are equally worried by the threat of waterlogging that the canal presents. They have calculated that
between 1,300 to 1,500 villages in Ropar, Samrala sub-division of Ludhiana district and Fatehgarh Sahib sub-division
of Patiala district will be waterlogged by the construction of
the SYL canal. 'Who is going to save us when our lands turn
saline?' asked Jarnail Singh, sarpanch of Doomcheri village.[28]

The farmers' point of view is supported by several experts. The Akali-Janata coalition government had planned
to dig the canal along the Shivalik foothills. A report submit-
ted by Dr H L Uppal, a water resource expert of the Punjab
Agricultural University in Ludhiana, had suggested that a
canal along the foothills would be more useful to farmers
and would create less ecological destruction. According to
Prof. Uppal, traditionally, Punjab's canals have followed
natural runoff patterns; they worked with nature's drainage
patterns. Large inter basin transfers have necessitated cross-
country flows across the contours of the regions. They work
against nature's drainage patterns. The Rajasthan Canal and
the SYL Canal are such cross-country channels. Such channels intercept the runoff of the tracts through which they
traverse. The cross drainage works cannot usually cope with
the obstructed runoff of the basin. This leads to problems
of waterlogging.

As Prof. Uppal has observed,

'Problems of waterlogging, salinity affecting agricul-
ture, roads, buildings, public health works, have been ex-
perienced in the south Punjab comprising Faridkot, Bhat-

inda, Muktsar, Kotkapura, Ferozepur, Fazilka, Jalalabad tracts which the cross country canals cross. Crores of rupees have already been spent on the drainage channels and are being spent on the construction of the additional drainage system in the tract to ameliorate the effect of the construction of the cross country canals and still the solution is not in sight.[29]

Prof. Uppal had anticipated similar problems with the SYL canal, which envisages transferring 6,500 cusecs of Sutlej River water into the Yamuna river system and 900 cusecs for the Punjab share of irrigation within the state. It runs along the left side of the Nangal Hydel channel to Ropar. From Ropar to the Haryana border at Kapoori, the canal has to run across the country.

As Prof. Uppal notes, 'the disadvantages in the case of this alignment are:

1. The catchment area on the north side of the alignments is considerable, and therefore, the runoff is high. Large drainage works shall have to be constructed. 100 cross drainage works would further disrupt the ecology. It is possible that flow from these works may effect the safety of the Bhakra canal.

2. On account of obstruction of the large runoff, waterlogging may occur on the north side of the SYLC.

3. The area between SYLC and Bhakra canal which is not very large would become land locked and drainage conjested.'[30]

Prof. Uppal had proposed the alternative alignment, along the Siwalik foothills, through Patiala-ki-Rao and Jainta Devi-ki-Roa. This alignment being higher to the north, would be effective in dealing with the runoff from the foothills and protecting the area below it against degradation. The hazards of waterlogging would be reduced, and farmers in the fertile tracts would not be dispossessed of their land. However, the suggestion of the alternative alignment was ignored, largely because of investments already made. Since the Irrigation Department had already installed turbines for generating 67 MW of power at the Nakia fall of the Anandpur Sahib Hydel project, which is linked to the SYL, any realignment would make the turbines redundant and seriously erode the utility of the Rs300-crore Anandpur Sahib project. They also pointed out that the present alignment, close to the old Bhakra canal, would be more expedient as it can draw on the available surveys of the ground water, bridges and rivulets channels. Yet even the officials accept that the worries of the Punjab farmers are genuine.

The Punjab-Haryana conflict over the SYL is, however, not merely a matter of a controversial alignment. It is part of the larger conflict of the sharing of river waters in a context of exploding demands for water. After two decades, the conflict is no longer merely over how the water should be shared, but also over how much water there is to share.

Distribution of water is a difficult task because water is a fluid and not a static resource. In the case of the Punjab rivers, it is not just the water, but the data on water that has been in a constant flux. In the 1955 Agreement, the surplus was assessed at 15.85 MAF. In the 1981 Agreement, this figure had been increased to 17.17 MAF. During the course of the arguments before the Eradi Tribunal, the Punjab government came to know that the irrigation experts deputed by the Tribunal had discovered 2 MAF of additional water flowing into the Punjab rivers which had not been taken into consideration earlier. Punjab had not confirmed this figure. This conflicting data-base has made the issue of water conflicts more complicated.

Apart from verifying the amount of water used on 1 July 1985, the tribunal had to decide the share of Punjab and Haryana from the remaining waters. It was also to verify the amount of water used by Rajasthan on that day but, as is evident from its terms of reference, Rajasthan's share would not be touched.

Punjab and Haryana both questioned each other's claims on the amount of water used on that particular day. According to the 1981 agreement between Punjab, Haryana and Rajasthan, Punjab was allotted 4.22 MAF, Haryana 3.50 MAF, Rajasthan 8.60 MAF. Delhi 0.20 MAF and Jammu and Kashmir 0.65 MAF out of the 17.17 MAF surplus from Ravi-Beas waters. However, according to the figures available at the Punjab irrigation department, Punjab was using 4.067 MAF and Haryana was using 1.331 MAF on 1 July, 1985. Converted into a commuted annual average, Punjab claims the use of 5.553 MAF. The Haryana Irrigation and Power Minister, Mr Shamsher Singh Surjewala challenged Punjab's claim, saying that it was using less than 2 MAF, while Haryana was using about 1.50 MAF and Rajasthan 3.50 MAF.

The total water available on that day was about 10 MAF. These claims are to be verified by the tribunal. However, it is the availability of total waters in a year that is important and it is here that there is considerable confusion. According to the 1981 agreement, the total surplus of Ravi-Beas waters is 17.17 MAF, after deducting the pre-Partition use of 3.13 MAF and transit loss in the Madhopur-Beas link of 0.25 MAF. This, however, is a mythical figure if the dependable availability is taken into account, which according to the Punjab irriga-tion department sources, averages about 11 MAF.

The figures of the Haryana irrigation department, on the other hand, show that the total availability in 1980-81 was 11.442 MAF and in 1981-82, 13.00 MAF. Out of 11.442 MAF, Punjab utilized 3.991 MAF against its share of 3.135 MAF, Haryana 1.672 MAF of its share of 2.267 MAF and Rajasthan 4.104 MAF against its share of 5.190 MAF. The downstream flow to Pakistan that year through Madhopur was 1.306 MAF and downstream to Ferozepur was 0.595 MAF. In 1982-83, Punjab utilized 3.843 MAF of its share of 3.599 MAF, Haryana 1.939 MAF out of 2.60 MAF and Rajasthan 4.951 MAF out of 5.955 MAF. In 1983-84, Punjab used 5.459 MAF, Haryana 1.433 MAF and Rajasthan 5.491 MAF.

It is thus clear that the amount of surplus water actually available is far less than 17.17 MAF, which was worked out in 1981. Even if the average dependable availability is 13 MAF and the 8.60 MAF share for Rajasthan is kept intact,

Table 4.9 : Statement Showing Shares and Utilization of Surplus Ravi-Beas Waters (in MAF)

1 Year		1980-81		1981-82
2. Surplus Ravi-Beas waters		11.442		13.004
3. Distribution	Share	Utilisation	Share	Utilization
(i) Punjab	3.135	3.991	3.599	3.843
(ii) Haryana	2.267	1.672	2.600	1.939
(iii) Rajasthan	5.190	4.104	5.955	4.951
(iv) J & K	0.650	0.300	0.650	0.300
(v) Delhi	0.200	0.370	0.200	0.428
Total:	11.442	10.437	13.004	11.461
4. Balance in the Reservoir	—	0.410	—	0.940
5. Wastage below Ferozepur	—	0.595	—	0.603
Grand Total:	11.442	11.442	13.004	13.004

there is virtually nothing left for Punjab and Haryana.[31]

Hence, from the Punjab viewpoint the need for the Sutlej-Yamuna Link Canal which is to carry the remaining share of Haryana from Punjab, must then be questioned. The Haryana leaders maintain that its share of 3.5 MAF will remain intact, which is the rationale behind the link canal project, eventhough in actual practice, 3.5 MAF is not available. Experts maintain that Haryana cannot utilize the canal to its full capacity which is 6.500 cusecs, even if the 1981 share is maintained. The water flow varies from year to year and Haryana's share ranges from 2.5 to 3 MAF.

New conflicts have emerged in Punjab because river flows are assumed to be static while they actually change over time, usually as a downtrend because of ecological degradation linked to water projects created for intensive irrigation and energy inputs for Green Revolution agriculture. The intensity of resource inputs creates an intense need for centralised control over water resources. The conflict over river waters in Punjab, is, in the final analysis, not merely a matter of sharing water between states, but of sharing power between the states and the centre. As Sant Longowal had stated in 1982, the Anandpur Sahib Resolution was aimed at reversing the trend of centralisation of power in India.

What the Punjab experience in conflicts over rivers has shown is that a 'just' distribution of water is not a matter of dividing a fixed stock of resources among a fixed set of needs, because neither the resources nor the demands are fixed. Injustice is being experienced by all parties concerned because of the fluidity of the resource, the exploding demand for water created by the Green Revolution and the fluidity of political actors making a demand on the resource and on its control. Water has become an important factor when it comes to the distribution of economic and political power in the Green Revolution areas.

Punjab has the largest area under irrigation in the country, and the spread of the Green Revolution has been concomittant with the spread of intensive use of water. Yet there is a feeling of loss and deprivation. Conflicts over the management and use of river waters is quite clearly a complex affair. Such conflicts are intensifying as the power of changing water distribution through water control technologies is increasing and with large projects, power is getting centralised.

The interference of the Central government in the control and management of Punjab waters increased after the Green Revolution became the model for agricultural development. New demands created new conflicts, and a new role was created for the Centre in the resolution of these conflicts. It is this process of centralisation that the Anandpur Sahib Resolution had challenged, even while it failed to challenge the development model which led to such centralisation.

River conflicts are increasing because new water projects are introducing drastic changes in river flows which can amount to the infringement of the territory (which is represented by the natural flow of the river) of other states. Also, these projects create an extension of a new concept of rights beyond the riparian rights determined by the natural flow of the river. While expanding the possible domain of beneficiaries, intensive irrigation simultaneously restricts and concentrates the availability of water to pockets of areas for intensive use. Non-riparian regions thus felt 'deprived' because every region has a new right to expect a share in intensive irrigation. The challenge posed by the Punjab crisis is a search for a concept of equity and conflict resolution in a paradoxical context of large scale manipulation of water flows which is based on a process of enclavisation, but which universalises the expectations in each region for becoming the beneficiaries of enclavised resource intensive development. In the case of enclavisation, limits to the sustainable and just use of resources are overlooked. Intensive irrigation involves the restriction of water use to small enclaves yet the shift from protective to intensive irrigation is justified by spreading the demand for more water as essential for agricultural development. Thus each region is drawn to the common expectation of becoming the beneficiaries of intensive development of an enclaved resource. It is within this paradox that the Punjab crisis creates the challenge for a new search for a concept of equity and conflict resolution.

References 4

1.   H L Uppal, *Water Resources of the Punjab : Their Potential, Utilisation and Management*, mimeo 1989.

2.   S Giriappa, *Water Use Efficiency in Agriculture*, Delhi: Oxford, 1983, p17.

3.   D S Kang, *op cit*, p200.

4.   H L Uppal and N S Mangat, 'Geohydrological Balance in Punjab', *Journal of the Institution of Engineers*, Vol.62, May 1982, pp365-371.

5.   Punjab Agricultural University, Department of Soils, mimeo, undated.

6.   Uppal and Mangat, *op cit*.

7.   Punjab Agricultural University, *op cit*.

8.   Punjab Agricultural University, *op cit*.

9.   Sidhu, *op cit*, p38.

10.   D S Kang, *op cit*, p201.

11.   C H Shah, *op cit*.

12.   'Beas, Sutlej project affects climate', *Indian Express*, Delhi, 18 August, 1986.

13.   'Satellite Studies show dam seepage', *Times of India*, Delhi, 1 April, 1989.

14. 'Breach causes water famine in Haryana', *India Express*, Delhi, 1 July, 1984.

15. 'Punjab Floods were Manmade', *Economic Times*, Bombay, 4 October, 1988. 'A friend brings sorrow', *Hindustan Times*, Delhi, 23 October, 1988. 'Dams and Floods', *Indian Express*, Delhi, 21 October, 1988.

16. 'Bhakra board chief shot dead', *Indian Express*, Delhi, 7 November, 1988.

17. Indian Law Institute , *Interstate Water Disputes in India*, Bombay: Tripathi, 1971.

18. *Ibid.*

19. 'River Waters Dispute', *Times of India*, Delhi, 15 & 16 November, 1985.

20. Pramod Kumar, *et al*, *Punjab Crisis: Context and Trends*, Centre for Research in Rural and Industrial Development, Chandigarh, 1984, p80.

21. *Ibid.*

22. 'The Anandpur Sahib Resolution', *Indian Express*, Delhi, 22 February, 1989.

23. 'Historic Accord with Akalis', *Indian Express*, Delhi, 24 July, 1985.

24. 'Rajasthan rejects Accord', *Indian Express*, Delhi, 26 July, 1985.

25. '29 Haryana MLA's resign', *Indian Express*, Delhi, 28 July, 1985. 'Call for Bhajan's resignation', *Indian Express*, Delhi, 24 March, 1985.

26. 'Dal Stand on Eradi report assailed', *Indian Express*, 5 February, 1987.

27. 'Sutlej Yamuna Canal, Jinxed Link', *India Today*, 30 November, 1985.

28. H L Uppal, 'Irrigation Canals Built in the plains of the Punjab – A review with special reference to their alignment in the context of the Sutlej Yamuna Canal', mimeo, 1987.

29. *Ibid.*

30. 'Tribunal to decide on Surplus water', *Times of India*, Delhi, 15 February, 1986. 'Decision on Sharing waters Difficult', *Indian Express*, Delhi, 15 February, 1986.

31. Pramod Kumar, *op cit*, p83.

5

THE POLITICAL AND CULTURAL COSTS OF THE GREEN REVOLUTION

THE ecological costs and natural resource conflicts associated with the Green Revolution were rooted in the replacement of cropping systems based on diversity and internal inputs with systems based on uniformity and external inputs. The shift from internal to externally purchased inputs did not merely change ecological processes of agriculture. It also changed the structure of social and political relationships, from those based on mutual (though asymmetric) obligations – within the village to relations of each cultivator directly with banks, seed and fertilizer agencies, food procurement agencies, and electricity and irrigation organisations. Further since all the externally supplied inputs were scarce, it set up conflict and competition over scarce resources, between classes, and between regions. Atomised and fragmented cultivators related directly to the state and the market. This generated on the one hand, an erosion of cultural norms and practices and on the other hand, it sowed the seeds of violence and conflict.

The centralised planning and allocation that made the Green Revolution possible in Punjab, affected people's lives

in direct ways. But it also affected their ideas of identity and
self. With government as referee, handing down decisions in
all matters, each frustration becomes a political issue. In a
context of diverse communities, that centralised control
leads to communal and regional conflict. Every policy decision is translated into the politics of 'we' and 'they'. 'We'
have been unjustly treated, while 'they' have gained privi-
leges unfairly. In Punjab, this polarised thinking gets expressed with the added dimension of religious discrimination against the Sikhs.[1]

The large scale experiment of the green revolution has
not only pushed nature to the verge of ecological breakdown, but also seems to have pushed society to the verge of
social breakdown.

In 1972, Francine Frankel in 'The Political Challenge of
the Green Revolution' had written that:

*'The green revolution is accompanied by an accelerated
disruption in traditional societies. More rapidly than in
other areas, traditional heirarchical arrangements rooted
in norms of mutual interdependence and (non-symmetric)
obligations give way to adversary relations based on new
notions of economic interest...*

*'It is not too early moreover, to consider one major impli-
cation of this analysis, namely that disruption is accelerated to so rapid a rate that the time available for autono-
mous re-equilibrating process, – even if such processes are
operative.....is critically curtailed. Thus, in the absence of
countervailing initiatives, forces already in motion will
push traditional societies in rural areas to a total breakdown.'* [2]

In 1972, the prediction of breakdown seemed far-fetched.

In 1989 it no longer seems so.

The rapid and large scale introduction of the Green Revolution technologies dislocated the social structure and political processes at two levels. It created growing disparities among classes, and it increased the commercialisation of social relations.

As Frankel observed, the Green Revolution was the instrument of a complete erosion of social forms. 'In those regions where the new technology has been most extensively applied, it has accomplished what a century of disruption under colonial rule failed to achieve, the virtual elimination of the stability residuum of traditional society – the recognition of mutual non-symmetric obligations by both the landed and landless classes.'[3]

While Frankel had predicted social breakdown, she had seen it as emerging from class conflict. Yet as the Green Revolution unfolded, it is the communal and ethnic aspects that come to the fore.

It was assumed that with development and modernisation communal conflicts will be swept away. However, recent experience suggests the opposite.

Modernization and economic development may, as in the case of Punjab harden ethnic identities and provoke or intensify conflict on the basis of religion, culture or race.

Most analysts of the crisis in Punjab have focussed exclusively on the politics of religion, or on the intrigue of electoral politics.[4,5] In this chapter I would like to go beyond the conventional reading of the conflicts in Punjab and trace how they are also linked structurally to the political, economic and cultural processes inherent to the Green Revolution transformation.

After an early experience of prosperity, Punjab farmers were rapidly disillusioned. In 1971-72, the returns on wheat cultivation were 27% on investment. By 1977-78 cultivators complained that their return had fallen to less than two percent of their investment. Even the well-to-do farmers started to experience the political and economic dislocation and indebtedness that the landless and smaller farmers had experienced immediately. After two decades of the rising debts and falling profits, the rich poor contradiction had become a centre-state conflict. Further, since Punjab farmers were Sikh farmers, and the regional party was the Akali Party, the Centre-State conflicts were quickly transformed into communal conflicts.

Three kinds of conflicts seem to have converged in creating what has been called the Punjab crisis.

The first is related to conflicts emerging from the very nature of the Green Revolution; such as conflicts over river waters, class conflict, the pauperisation of the lower peasantry, the use of labour-displacing mechanisation, the decline in the profitability of modern agriculture etc., all heading to a disaffected peasantry engaged in farmers' protests.

Secondly, there were conflicts related to religion-cultural factors and revolving around the Sikh identity. These conflicts were rooted in the cultural erosion of the Green Revolution which commercialized all relations, and created an ethical vacuum where nothing is sacred and everything has a price. Religious revivalism which emerged to correct the moral and social crisis crystallised finally in the emergence of a separatist Sikh identity.

The third set of conflicts were related to the sharing of

economic and political power between the centre and state.

The shift from local organisation and internal inputs to centralised control and external and imported inputs restricted the nature of power between the farmers and the government and the state and the centre. Just as at the level of natural resources, the shift from diversity to monocultures, and the shift from internal inputs to external inputs of seeds and chemicals led to ecological vulnerability of agricultural ecosystems, the associated shift to external dependence politically led to societal vulnerability.

The rise of the market and rise of the state that was part of the Green Revolution policy led to the destruction of community and the homogenising of social relations on purely commercial criteria. The shift from internal farm inputs to centrally controlled external inputs shifted the axis of political power and social relations. It involved a shift from mutual obligations within the community to electoral politics aimed at state power for addressing local agricultural issues.

The processes of centralisation were associated with the processes of homogenisation. To a large extent the movements for regional, religious, and ethnic revival are movements for the recovery of diversity in a context of homogenisation. The paradox of separatism is that it is a search for diversity within a framework of uniformity, it is a search for identity in a structure based on erasure and erosion of identities. The shift from Sikh farmers demands to the demand for a separate Sikh state comes from the collapse of horizontally organised diverse communities into atomised individuals linked vertically to state power through electoral politics. The ecological crisis of the Green Revolution is thus mirrored in a cultural crisis caused by an erosion of diversity and structures of local governance and the emergence of

homogenisation and centralised external control over the
daily activities of agriculture food production.

The Economic Costs: A narrow and shortlived prosperity

By the 1980s, the optimism of the Green Revolution had
faded in Punjab. The farmers, rich and poor alike, were
feeling the pinch of ecological erosion, debt and declining
profit margins. In addition, they were beginning to react to
the cultural erosion that had been the result of the spread
of commercial agriculture. Finally, since the policies in which
they were trapped were policies created by the Centre, they
felt a sense of exclusion from meaningful political participation in key decisions that affected their socio-economic
status. Farmers' protests and agitations around these issues
therefore came to the centre stage of Punjab politics, in the
early 1980s. By the mid-1980s however, politics in Punjab
was totally communalised, and conflicts emerging from an
agricultural crisis were rendered intractable.

The Green Revolution was primarily a technocratic response to problems of growing conflict in agrarian societies.
However, soon after the introduction of the new technologies, it was being recognised that they could generate a new
range of agrarian conflicts, by increasing the polarisation
between rich and poor farmers. A 1969 report of the Home
Ministry on 'The Causes and Nature of the Current Agrarian Tension' identified inequality in landholdings as a 'pre-
disposing' factor for agrarian tensions. However, the 'proxi-
mate' causes for open conflict were located in the new agricultural strategy and the Green Revolution.

The inequality generating effects of the Green Revolution were built into the strategy of 'building on the best' – the best endowed region and the best endowed farmers. The increase in resource intensity of inputs for Green Revolution agriculture implied the increase in capital intensity of farming which tended to generate new inequalities between those who could use the new technology profitably, and those for whom it turned into an instrument of dispossession. The poorer peasants were unable to maintain their land holdings under the high input economies of the Green Revolution. Between 1970 and 1980, a large number of smallholdings disappeared in Punjab due to economic non-viability. In 1970-71, the total number of operational holdings in Punjab was 1,375,382 which fell to 1,027,127 in 1980-81, a decline of nearly 25%.[6]

Dasgupta has produced evidence which points to the conclusion that in Punjab 'the distribution of operated land has shifted in favour of the richer farmers under the new technology'.[7] According to Bhalla, both in the high yielding wheat and rice area, 'the distribution of operated land has shifted in favour of big farmers'.[8]

The Green Revolution thus started a process of depeasantisation of the peasantry, through increasing costs of cultivation. Agricultural labour also does not seem to have gained from the Green Revolution in Punjab. Bardhan concludes that 'even in the prime' Green Revolution area of Punjab/Haryana, the proportion of people below a bare minimum level of living has increased.[9] A recent comprehensive compilation of data for the Punjab, covering the years between 1961 and 1977, (The data provided indexes of real wage rates for each year, for the whole state and district on ploughing, sowing, weeding, harvesting, cotton picking, and other agricultural operations) revealed that 'in many years the rise in money wages lagged behind price changes, leading to reduced real wage rates for most operations between 1965 and 1968, and again in 1974, 1975 and 1977'.[10]

Class conflict emerging from the polarisation of rural society has been an old feature of rural India. In recent years, the rural/urban conflict has become a new preoccupation of farmers' movements, which see rural 'Bharat' as exploited by the needs of the urban elites for cheap food and raw material. The Green Revolution strategy, was in fact a strategy for creating cheap food surpluses for the growing urban/industrial centres. In the early years, food subsidies and support prices created an artificially profitable economic package for the Punjab farmers, especially the more prosperous ones. However, the intensive input agriculture needed credit, which over time was converted into indebtedness. Further, costs of inputs kept rising as higher rates of fertiliser and pesticides had to be applied to maintain yields. In addition, the high subsidies and support prices of the early years could not be maintained indefinitely. Thus while

Table 5.1: Procurement price, cost of production and rate of return over cost in wheat cultivation in Punjab

Year	Procurement price (Rs. per quintal)	Cost of production (Rs. per quintal) (percent)	Rate of return over cost
1970-71	76	61.04	24.50
1971-72	76	59.71	27.28
1972-73	76	67.10	13.26
1973-74	76	74.34	2.23
1975-76	105+8*	99.45	13.62
1976-77	105	101.39	3.56
1977-78	110	108.57	1.32
1978-79	112.50	101.45	10.89

*Bonus

Source: *Rajbans Kaur, 'Agricultural Price Policy in Developing Countries with special reference, to India' unpublished PhD Thesis, Punjab University, Patiala, 1982, p275.*

in the initial phase of the Green Revolution, agriculture was a paying proposition with high rates of return over cost, it had very rapidly created a crisis of indebtedness and falling rates of return. In Punjab, the average excess of procurement price over cost of production changed from (-) 14.0% during 1954-55 through 1956-57 to (+) 124.5% in 1970-71. Over time, the subsidies that had made agriculture artificially remunerative, were reduced.[11] Table 5.1 shows how the rate of return over cost in Punjab declined from 25.89% on average for 1970-71 and 1971-72 to an average of 6.11% during 1977-78 and 1978-79.

The burden of declining rates of return was strongest among small farmers. Small farmers with land size of 5 acres or less constitute 48.5% of the cultivating households in Punjab. According to a survey, in 1974 small farmers were annually running a per capita loss of Rs125 whereas farmers with land between 5 and 10 acres were producing a per capita profit of Rs50 while farmers with land more than 20 acres were producing a per capita profit of Rs1,200. Another survey carried out between 1976-77 and 1977-78 indicates that marginal and small farmer's households were annually running into an average deficit of Rs1512.17 and Rs1648.19. Another study points out that 24% of small farmers and 31% of marginal farmers live below the poverty line in the Green Revolution state of Punjab.[12] The Johl Committee report also confirms that except for an increase in per hectare income during 1977-78 and 1978-79, there has been a decline in returns from farming in Punjab. Even during the period when agriculture in the State witnessed an overall fast rate of growth, farm incomes in real terms have shown a stagnation from the early eighties. There is, in fact, even evidence of a decline in the real income per hectare from 1978-79 onwards. According to Johl, this tendency is expected to get accentuated because stagnation had been reached at all levels – area cropped, prices, and productivity of the two

principal crops viz. paddy and wheat which had all reached
their limits by the early 1980s.

The short term economic viability of high profits for the
'progressive' farmer of Punjab, and cheap and assured food
supplies for India's urban population was created at what
was to eventually be a high political cost for Punjab's farmers. Green Revolution agriculture of intensive inputs was
made possible with the introduction of agricultural credit.
The Agricultural Refinance Corporation (ARC), a consor-
tium of commercial banks was set up in 1963 to give medium
and long term credit for major macro-level development
projects. ARC has since been reconstituted as NABARD, the
National Bank for Agriculture and Rural Development. The
creation of artificial profitability for the production of Green
Revolution wheat and rice was based on the creation of
highly centralised institutions for the control of farm economies. Two central bodies related to food production, procurement and distribution were established in 1965 on World
Bank advice. One was the Food Corporation of India (FCI)
which was responsible for procurement, import, distribution, storage and the sale of food grain. The other was the
Agricultural Prices Commission (APC) which determined
the minimum support prices for food grains, and through it,
controlled cropping patterns, land use and profitability.
Through food price and procurement, the Central government now controlled the economics of food grain production and distribution. The profitability of food-grain production in this centralised and enclavised form could not be
maintained overtime. The economics of the artificially high
procurement price of Rs76 per quintal in the early days of
the Green Revolution, which was nearly Rs15 per quintal
higher than market prices, was based on compensating the
high costs of domestically procured foodgrain and pooling
them with the low priced imported foodgrains. With the
phasing out of concessional imports of foodgrains in the

1970s, it became impossible for the Central government to maintain high procurement prices without heavy losses. In addition, the intensive agriculture approach which had focussed on the best endowed regions like Punjab, lowered the food producing and food purchasing capacities of other regions which were resource poor and had been excluded from the new agricultural strategy.

The concentration on paddy and wheat in Punjab agriculture, and the lack of purchasing power in the neglected regions of rural India, led to a build up of heavy surpluses which Punjab could not sell profitably. Meanwhile, food deprivation was increasing largely as a result of the Green Revolution strategy. The 70% of agricultural land in the country left out of the intensive development suffered decline in productivity due to neglect and lack of inputs. People from these regions grew poorer.

The imbalance was not just regional, but also related to crops. The increase in rice and wheat production in Punjab had been, in part, achieved by creating a scarcity in oilseeds and pulses, necessary for a nutritional balance in vegetarian diets. The logic of specialisation – regionwise and cropwise – and the related logic of profitability of the Green Revolution strategy reached its limits in the 1980s. During the 1985 paddy marketing season, the farmers of Punjab could not sell their produce profitably. The problem started with the late announcement of a procurement price for paddy, as a result of which the procurement agencies started their produce at very low prices to rice millers and traders. The crisis in marketing of paddy made it evident to the farmers and the government of Punjab that they had to break out of centrally controlled specialised production of food grain and go in for diversification. A committee, under the Chairmanship of S S Johl was thus formed to advise the Punjab government on the diversification of Punjab's agriculture, to free it from centralised control.

In less than two decades. Punjab's farmers, the main beneficiaries of the Green Revolution strategy were beginning to feel victimised. Punjab's farm organisations were appealing to the farmers to free themselves of the new colonialism. Since it was the agencies of the Central government that controlled the policies related to the new agriculture, the conflict that emerged in Punjab became primarily a conflict between Punjab and the Central government. The unique role that the Green Revolution strategy had created for Punjab for being the bread basket of India also became the reason for new discontent, when the benefits started fading. By the 1980s, Punjab's farmers were organising themselves on the grounds of being treated like a colony of the Centre to feed India. 'For the past three years, we have increasingly lost money from sowing all our acreage with wheat. We have been held hostage to feed the rest of India. We are determined that this will change'.[13]

The 1980s were marked by farmers agitations over the high costs of agricultural inputs. Campaigns were launched on water rates, electricity rates, procurement prices etc. On 31 January 1984, a call for 'rasta roko' (road blockade) was given and farmers got Rs12.5 crore in relief for the damage caused by pests to their cotton crop. On 12 March 1984, the Bhartiya Kisan Union started a gherao (blockade) of the Punjab Raj Bhawan (Governor's residence) at Chandigarh demanding a stop to higher electricity rates, higher procurement price for wheat and the scraping of APC and its replacement by an agricultural 'cost' commission. Finally on 18 March 1984, an agreement was reached and the gherao was lifted.

However, since the farmers' grievances arising from a nonsustainable miracle remained, farmers' agitations did not stop. During April 1984, the Bharatiya Kisan Union (BKU) focussed its campaign on indebtedness, which was called 'karja roke'. In most villages in Punjab, the BKU hung notices at village entrances and crossings which said, 'without proper accounting, recovery of loans is illegal. Entry of recovery staff in the village is not allowed without permission. By order, BKU'. The high dependence on credit necessary for high inputs and the falling rates of return on investment had left most Punjab farmers heavily indebted. The agricultural credit in Punjab works out at Rs103 per hectare for short term production credit alone, against a national average of Rs35. Squeezed by debts and declining returns, Punjab's farmers were protesting throughout the state during the first half of 1984.

Communalising the Farmers' Protests

In May 1984, the farmers' agitation was at its height in Punjab. For a week, from 10 to 18 May, farmers gheraoed the Punjab Raj Bhawan (Governor's House) since the state was under the President's rule. By conservative estimates, at any time, more than 15-20 thousand farmers were present in Chandigarh during the gherao. Earlier, from 1 to 7 May, the farmers had decided to boycott the grain markets to register their protest against the Central government procurement policy. On 23 May 1984, Harchand Singh Longowal, the Akali Dal President, announced that the next phase of the agitation would include attempts to stop the sale of food grain to the Food Corporation of India. Since Punjab provides the bulk of the reserves of grain, which are used to sustain the government distribution system and thus keep prices down, a successful grain blockade implied a serious national crisis and would have given Punjab a powerful bargaining tool for its demands for greater state autonomy. On 3 June, Mrs Gandhi called out the army in Punjab and on 5th of June the Golden Temple was attacked, which was for the Sikhs, an attack on the Sikh faith and Sikh dignity and honour. After Operation Bluestar, as the military operation was called, the Sikhs as a farming community has been forgotten; only Sikhs as a religious community remain in the national consciousness. Nothing after that could be read without the 'communal' stamp on it. The movement in Punjab was no longer primarily a movement to protect the farmers' interests but had been submerged by a movement to protect the Sikh identity.

A crisis that was economically and politically rooted in the Green Revolution rapidly expressed itself in communal overtones because of the contingent overlapping of the identity of the farming community in Punjab with a Jat Sikh identity. Discontent that had expressed itself in Punjab in the 1980's was the result of centrally controlled agricultural production and the resulting economic and political crisis. It was located in Centre-state politics and the political economy of the Green Revolution. However, because of the contingent factor that farmers in Punjab are largely Jat Sikhs, and the party representing their interests is the Akali party, it was possible to represent Green Revolution related conflicts as communal conflicts, and treat them as only having a religious base unrelated to the politics of technological change and its socio-economic impact. The Green Revolution failed to bring lasting peace and prosperity to Punjab. Instead, it sowed the seeds of violence and discontent, which got an increasingly communal colouring as a source of diversion from the politics of the Green Revolution. There were, of course, cultural reasons which facilitated the communalising of the Punjab crisis. The Green Revolution package was not just a technological and political strategy. It was also a cultural strategy which replaced traditional peasant values of co-operation with competition, of prudent living with conspicuous consumption, of soil and crop husbandry with the calculus of subsidies, profits and remunerative prices.

The conflicts arising at the ecological and economic levels converged with the cultural conflicts between traditional values and the culture of conspicuous consumption created by the short term affluence and profitability of the Green Revolution. The 1980s saw in Punjab a genuine cultural upsurge as a corrective to the commercialisation of Punjab's culture. Two decades of rapid transformation of the economy, society and culture of Punjab had generated an ethical and moral crisis. The overriding culture of cash and profitability disrupted old socialties and fractured the moral norms that had governed society. Circulation of new cash in a society whose old forms of life had been dislocated led to an epidemic of social diseases like alcoholism, smoking, drug-addiction, the spread of pornographic films and literature and violence against women.

Religion provided a corrective source of values to this cultural degeneration, and was also a source of solace to the victims of the violence that was associated with the new forms of degenerate consumption. Sant Jarnail Singh Bindranwale, who later became the leading ideologue for separatism and religious fundamentalism, gained his early popularity with the Punjab peasantry by launching an ideological crusade against the cultural corruption of Punjab. The most ardent followers of Bhindranwale in his first phase of rising popularity were children and women, both because they were relatively free of the new culture of degenerative consumption, and they were worst hit by the violence it generated. In the second phase of Bhindranwale's popularity, men also joined his following, replacing vulgar movies with visits to gurdwaras, and reading the 'gurbani' (teachings of the Gurus) in place of pornographic literature. The Sant's following grew as he successfully regenerated the 'good' life of purity, dedication and hard work by reviving these fundamental values of the Sikh religion. The popularity of Bhindranwale in the minds of the Sikhs in the countryside was based on this positive sense of fundamentalism as revitalising the basic moral values of life which had been the first casualty of commercial capitalism. During the entire early phase of Bhindranwale's preachings, he made no anti-government or anti-Hindu statement, but focussed on the positive values of the Sikh religion. His role was largely that of a social religious reformer.

Sikh revivalism took a negative turn and became nega-tivistic-fundamentalism with the interference of party poli-tics, especially from the Congress(I).[14] Bhindranwale was used against the Akalis, for electoral gains and when he was himself killed in the Golden Temple after Operation Blues-tar, Punjab politics and the Sikh religion had been totally communalised. After Operation Bluestar and the Anti-Sikh Violence in November 1984, following Indira Gandhi's assassination, the issue of the danger to the Sikh identity became a central concern in Punjab. The communal form of Sikh revivalism came to dominate over the earlier regenera-tive and ethical form. However, it never became a large scale communal conflict between Hindus and Sikhs within Punjab, but was and continues to be a conflict between what is perceived as a communal Hindu Centre and a Sikh people in a Sikh state.

The 'Gurmata' (Collective resolution by the Congrega-tion) passed at a Sarbat Khalsa (All Sikh Convention) on 13 April, 1986 expresses this perception of the communal con-flict as primarily a Centre-state conflict explicitly.

'If the hard-earned income of the people or the natural re-sources of any nation or region are forcibly plundered; the goods produced by them are paid at arbitrarily deter-mined prices while the goods bought by them are sold at

high prices and in order to carry this process of economic exploitation to its logical conclusion, the human rights of people or of a nation are crushed, then these are the indices of slavery of that nation, region or people. Today, the Sikhs are shackled by the chains of slavery. This type of slavery is thrust upon the states and 80% of India's population of poor people and minorities. To smash these chains of slavery, Sikhs, on a large scale, by resorting to reasoning and by using force and by carrying along with them these 80% people of India, have to defeat the communal Brahmin-Bania combine that controls the Delhi Durbar. This, is the only way of establishing hegemony of Sikhism in this country. In this way, under the hegemony of Sikh world-view and politics, a militant organisation of the workers, the poor, the backward people and the minorities (Muslims, Christians, Buddhists and Dalits etc.) has to be established.' [15]*

In all the Gurmatas, the hold of money-power in social life is decried and this is considered the cause of the deterioration in the moral fibre of society and the cause of the rise of greed and selfishness and their consequences in social and political life. The concentration of political power is also subjected to sharp criticism and the necessity for collective democratic decision making is emphasised. The regeneration of Sikhism is linked to the regeneration of tradition, 'to liberate oneself from money, ego, cowardice, ignorance, arrogance and stupidity and instead of seeking collective welfare through one's own welfare, to seek one's welfare through the collective welfare.'[16]

From the foregoing analysis it becomes clear that the revival of the cultural identity of the Sikhs was more in response to the erosion of regional autonomy and the cultural and moral erosion of life in Punjab by the commercial culture of the Green Revolution. It was not a cultural conflict of Sikhs with people of the Hindu religion. In 1978, at the time of the Anandpur Sahib resolution, issues of river waters, food production and pricing, and the political economy of Centre-state relations were at the top of the demands. The Centre systematically avoided these political and economic issues and communalised the situation after the 1980 elections by calling the Anandpur Sahib resolution secessionist and by withdrawing the case that the Akalis had filed on water disputes in 1978 in the Supreme Court. By avoiding the regional and developmental issues raised by the Anandpur Sahib Resolution, the Centre avoided responding to a serious developmental crisis. Instead of responding to the valid demand for redefining Centre-state relationships, it raised the bogey of Sikh separatism. The Dharam Yudh declared by the Akali Dal on 4 August 1982, has to be interpreted in the Indian meaning of Dharam as justice and rights, not as religion. Dharam Yudh is a fight for justice, not a religious war against members of another religious community. Dharam Yudh became central to Punjab after Operation Bluestar and the November 1984 Sikh massacres because over and above the violation of regional autonomy, the people of Punjab were also victims of a violation of cultural and religious identity. Finally, all these violations relate to issues of rights and justice, and the violator of these rights was not another ethnic or religious group, but the centre of power of a national security state.

Since Independence, India has constantly been embroiled in tensions between the Centre and the federating units. This has led political analysts to assume that the cultural diversity in India will soon bring down the fragile unity in the Centre which embodies the nation state. However a closer inspection of earlier conflicts between the Centre and the federating units reveals that cultural factors *per se* have never put any tension on the political unity of this country. If anything, purely cultural assertions have been easily absorbed by the Indian nation state because its power of decision making was increased. The first of these was the demand for unilingual states. During the linguistic movements the Centre had to preside over the reshaping of the administrative structure and boundaries of the constituent elements but it did not have to deny the linguistic demands of any major linguistic group by accepting the linguistic demands of another. Its own power increased, without decreasing the power of constituent units. But with regional movements emerging from conflicts related to the developmental process, the ruling party at the centre, more particularly the Congress (I), was for the first time faced with a political formation that was in conflict with it on political and economic grounds. To give in to the new demands was to give up the immense power the Centre had acquired through restructuring agricultural policy. It therefore had to divert the politics.

As observed by a commentator,

> 'now was the time for the Centre to strike back. And the Centre struck back ethnically. It ethnicised secular issues in order to marginalise its opponents, one by one, from the national main stream. The fact that the demarcation of state boundaries is superimposed by linguistic and/or religious markers provided the temptation for the regional political formations to lapse into the ethnic slot the Congress was pushing them into.' [17]

Development, Social Disintegration and Violence

The process of development leads, in effect, to turning one's back to the soil as a source of meaning and survival, and turning to the state and its resources for both. The destruction of organic links with the soil also leads to destruction of organic links within society. Diverse communities, co-operating with each other and the land become different communities competing with each other for the conquest of the land. The homogenisation processes of development do not fully wipe away differences. Differences persist, not in an integrating context of plurality, but in the fragmenting context of homogenization. Positive pluralities give way to negative dualities, each in competition with every 'other', contesting for the scarce resources that define economic and political power. The project of development is unleashed as a source of growth and abundance. Yet by destroying the abundance that comes from the soil and replacing it with resources of the state, new scarcities and new conflicts for scarce resources are created. Scarcity, not abundance, characterizes situations where nothing is sacred but everything has a price. As meaning and identity shifts from the soil to the state and from plural histories to a singular history of the Rostowian path, ethnic, religious and regional differences which persist are forced into the straight-jacket of 'narrow nationalism'. Instead of being rooted spiritually in the soil and the earth, uprooted communities root themselves in models of power presented by the nation state. Diversity is mutated into duality, into the experience of exclusion, of being 'in' or 'out'. The intolerance of diversity becomes a new social disease, leaving communities vulnerable to breakdown and violence, decay and destruction. The intolerance of diversity and the persistence of cultural differences sets up one community against another in a context created by a homogenising state, carrying out a homogenising project of development. Difference, instead of leading to richness of diversity, becomes the base for diversion and an ideology of separatism.

In the South Asian region, the most 'successful' experiments in economic growth and development have become, in less than two decades, crucibles of violence and civil war. Culturally diverse societies, engineered to fit into models of development have lost their organic community identity. From their fractured, fragmented and false identities, they struggle to compete for a place in the only social space that remains – the social space defined by the modern state.

The upsurge of ethnic religious and regional conflicts in the Third World today may not be totally disconnected from the ecological and cultural uprooting of people, deprived identities, pushed into a negative sense of self with respect to every 'other'. Punjab, the exemplar of the Green Revolution miracle until recently one of the fastest growing agricultural regions of the world is today a region riddled with conflict and violence. At least 15,000 people have lost their lives in Punjab in the last six years. During 1986, 598 people were killed in violent conflicts. In 1987, the number was 1,544. In 1988 it had escalated to 3,000. And 1989 shows no sign of peace in Punjab. Punjab is the most advanced example of the disruption of links between the soil and society. The Green Revolution strategy integrated Third World farmers into the global markets of fertilizers, pesticides and seeds, and disintegrated their organic links with their soils and communities. The progressive farmer of Punjab became the farmer who could most rapidly forget the ways of the soil and learn the ways of the market. One outcome of this was violence to the soil resulting in water logged or salinated deserts, diseased soils and pest-infested monocultures. Another outcome was violence in the community, especially to women and children. Commercialisation linked with cultural disintegration created new forms of addictions and new forms of abuse and aggression.

The religious resurgence of the Sikhs that took place in the early 1980s was an expression of a search for identity in the ethical and cultural vacuum that had been created by destroying all value except that which serves the market place. Women were the most active members of this movement of resurgence. There was also a parallel movement of farmers, most of whom happened to be Sikhs, protesting against centralised and centralising farm policies of the state which left the Punjab farmers disillusioned after a short-lived prosperity. The struggles of Sikhs as farmers and as a religious community were, however, rapidly communalized and militarized. On the one hand, the people of Punjab became victims of state terrorism exemplified by the attack on the most sacred shrine of the Sikhs – the Golden Temple. On the other hand, they were victims of the terrorism of Sikh youth whose sense of justice was constrained by the political contours of a narrow state concept of the Sikh identity. Punjab, the land of the five rivers, was forgotten and redefined as Khalistan. The soil gave way to the state as the metaphor for organising the life of society.

The conflicts were thus relocated in a communalised zone for the contest of statehood and state power. They moved away from their beginnings in tension between a disillusioned, discontented, and disintegrating farming community and a centralising state which controls agricultural policy, finance, credit, inputs and pieces of agricultural commodities. And they also moved away from the cultural and ethical reappraisal of the social and economic impact of the Green Revolution.

The Green Revolution was to have been a strategy for peace and abundance. Today there is no peace in Punjab. There is also no peace with the soils of Punjab and without that peace, there can be no lasting abundance.

The communalising of Punjab has been the result of a process of political and economic transformation of a region to create the 'miracle' of the Green Revolution. Instead of generating peace, this strategy of transformation has generated violence and bloodshed. And as one technological fix fades away, a second Green Revolution is offered as a cure for the political and economic problems inherited from the first.

References 5

1. Jeffrey Robin, *What is happening to India?*, London: Macmillan, 1986, p37.

2. Francine Frankel, *The Political Challenge of the Green Revolution*, Centre for International Studies, Princeton University, 1972, p38.

3. *Ibid*, p4.

4. Mark Tully and Satish Jacob, *Amritsar: Mrs Gandhi Last Battle*, Delhi: Rupa Publishers, 1985.

5. Rajiv A Kapur, 'Sikh Separatism', *The Politics of Faith*, Herts (UK): Allen and Unwin, 1986.

6. S S Gill, 'Contradictions of Punjab Model of Growth and Search for an alternative', *Economic and Political Weekly*, 15 October, 1988.

7. Biplab Dasgupta, *Agrarian Change and the New Technology in India*, Geneva: UNRISD, 1977, pp162-64, 167.

8. G S Bhalla, *Changing Structure of Agriculture in Haryana*, A study of the Impact of the Green Revolution, Chandigarh: Punjab University, 1972, pp269-85.

9. Kalpana Bardhan, 'Rural Employment, Wages & Labour Markets in India', *Economic and Political Weekly 2*, No.27, 2 July, 1977, pp1062-63.

10. Sheila Bhalla, 'Real Wage Rates of Agricultural Labourers in Punjab 1961–77', *Economic and Political Weekly 14*, No.26, 30 June, 1979.

11. S S Gill and K C Singhal Farmers, 'Agitation Response to Development Crisis of Agriculture', *Economic and Political Weekly*, 6 October, 1984.

12. Gopal Singh, *Socio-Economic Basis of the Punjab Crisis*, Vol. XIX, No.l 7 January, 1984, p42.

13. The statement was made by a representative of a Punjab farm Organization, *Christian Science Monitor*, 30 May, 1984, p10.

14. Mark Tully and Satish Jacob, *Amritsar: Mrs. Gandhi's Last Battle*, Delhi: Rupa Publishers, 1988.

15. Pritam Singh, *Two facets of religious revivalism: A Marxist viewpoint of the Punjab question*, Punjab University, mimeo.

16. *Ibid*.

17. Dipankar Gupta, 'Communalising of Punjab – 1980-85', *Economic and Political Weekly*, Vol XX, No.28, 13 July, 1985, p1185.

Today, Punjab has the promise of a hundred years of spring.

Today, the day of Baisakhi, the harvest festival, H.E. Shri Siddhartha Sankar Ray, the Governor of Punjab, will lay a foundation stone at village Channo in the district of Sangrur. This stone will grow into a future of plenty for the farmers of Punjab. The Pepsico-Voltas-Punjab Agro venture will generate ideas to grow better fruits and vegetables and create an eager marketplace for our high quality produce. On this harvest festival, together we will sow the seeds of prosperity.

PEPSI FOODS LIMITED

6

PEPSICO FOR PEACE?
The Ecological and Political Risks of the Biotechnology Revolution

THE PUNJAB crisis, characterised by violence and discontent, is in large measure linked to the unanticipated effects of the technological fix of the Green Revolution which was ironically aimed at preventing violence and containing discontent through the technological transformation of Indian agriculture. The main elements of the technological fix of the first Green Revolution were:

1. the replacement of diverse mixed crops and rotational cropping patterns of cereals, pulses and oilseeds produced primarily for self-consumption, with monocultures of introduced wheat and rice varieties produced primarily for the market.

2. the substitution of internal resources of the farm with purchased inputs of seeds, fertilisers, pesticides, energy, etc.

3. the enclavisation of food production for the entire country in a small region.

The Green Revolution has been projected as having

increased agricultural productivity in an absolute sense. At the level of resource utilisation, the new seed-fertilizer technology was clearly counter productive, both with respect to natural resource inputs like water, as well as industrial inputs such as fertilizer use. This is clear in the case of rice where some indigenous high yielding varieties compared with the yields of the Green Revolution varieties, but used substantially less water and fertilizer inputs. Richaria reported yields of 4,000 kg and more in rice cultivation in Baster,[1] and Yegna Iyengar recorded rice yields above 5,000 kg/ha in South India.[2] Even in the case of wheat, and particularly true in Punjab, increase in yields was achieved; the increase in cost of inputs did not affect the gain in yields. Productivity with respect to water use and fertilizer use thus actually declined, as summarised in the Table below.

Table 6.1: Comparison of productivity of native varieties and Borlaug varieties of wheat

	Native variety	Borlaug variety
Yield Kg/ha	3291	4690[3]
Water Demand	12"	36"[4]
	5.3 cm	16 cm
Fertilizer Demand	47.3	88.5[5]
Productivity with respect to water use (kg/ha/cm)	620.94	293.l
Productivity with respect to fertilizer use (kg/ha/kg)	69.5	52.99

In terms of energy too, the Green Revolution technology is far more inefficient than the technologies it displaced. When food energy output from rice-growing systems is compared with the total energy input, the energy use in agriculture is found to have declined. The pre-Green Revolution systems have energy ratios of around l0, but with the introduction of the Green Revolution this level is more than

halved. With industrial agriculture, the ratio is further reduced to one, and as much energy is being expended in
these systems as is being obtained from them in the form of
food.[6]

While the increase in productivity was the primary objective of the Green Revolution, in terms of resources and
energy, the productivity actually declined. The increase that
was achieved in the early phases was at the level of financial
returns. In fact, the motive force for the Green Revolution
technology package came from profits – profits for agribusiness, and profits for farmers. However, the ecology of the
Green Revolution demanded increasing costs of inputs and
resulted in decreasing profits for the farmers of Punjab.
Agricultural income stagnated or began to decline. In less
than two decades, the Green Revolution had become financially and ecologically unviable, though it succeeded in the
production of surpluses of specialised crops in a specialised
region for a short period, based on high inputs, high subsidies and high support prices.

There are two options available for getting out of the
crisis of food production in Punjab and making agriculture
economically viable again. The first is to move away from
resource and capital intensive agricultural technology to
low-cost agriculture by making food production economically and ecologically more viable again, through the reduction of input costs. The second option is to move away from
staple foods for domestic markets to luxury foods and non-
food crops for export markets, with a new dependence on
imports of high-technology inputs like seeds and chemicals.
It is the latter option that has been officially adopted as
the strategy for the second agricultural revolution in Punjab.

The main elements of the technological fix of the second
Green Revolution are:

1. the substitution of wheat and rice produced for
domestic markets with fruits and vegetables produced for
the export of processed foods.

2. the substitution of Green Revolution technologies
with new bio-technologies, integrated more deeply with
farm chemicals on the one hand and food processing on
the other.

3. the total neglect of staple food production as a primary objective of public policy.

Pepsico for Peace?

The Pepsico project to be located in Punjab is at the
Centre of the new agriculture policy. It has been viewed by
commentators as 'a catalyst for the next agricultural revolution', and a 'programme for peace'.[7] The first agricultural
revolution was also to have been a strategy for peace and
prosperity. It delivered violence and discontent. Will the
second agricultural revolution in Punjab succeed where the
first failed?

The Pepsico project now called Pepsi Foods, was first
proposed in 1986 as a collaboration between Punjab Agro-
Industries Corporation, Voltas, a subsidiary of the Tata's,
and Pepsico, the US multinational. The project consists of
four activities, namely (1) an agro-research centre to develop improved varieties of seeds using biotechnology, (2) a
potato and grain-based processing plant to produce high-
quality food products, (3) a fruit and vegetable processing
plant for processing fruits, (4) and a soft drink concentrate
unit to make soft drink and juice concentrates. The cost of
the project was initially Rs21 crore, of which Pepsico's share
was about Rs3.59 crore. The project envisaged an export of
Rs20 crore in the first year alone. The total value of exports
over ten years was expected to be Rs194 crores. About 74%

of the total outlay of Rs22 crores on the project was in the processed food sector The plan was to utilize one lakh tonnes of fruit and vegetables, which will be grown on land that now grows cereals. The import component of the project was estimated at Rs37 crores over a period of 10 years. In March 1989, the partners in the Pepsico project announced a doubling of their investment. They now plan to invest more than Rs50 crores in the project.[8]

When inspite of two years of controversy and debate, the Central government cleared the Pepsi project on 19 September 1988, the minister of the newly-formed Food Processing Industry Ministry, justified it on the grounds of diversification of agriculture, increase in agricultural income and employment, and restoration of peace and stability in Punjab.[9] The project has been criticized for creating a new dependency, sacrificing self-reliance and going against the national interest.[10] Like the Green Revolution in Punjab, the Pepsico 'programme for peace' claims to offer a technological fix for a political crisis. Like the Green Revolution before it, it holds the potential to aggravate the crisis by introducing new vulnerability in agriculture.

Seeds of Ecological Vulnerability

The 'miracle seeds' of the Green Revolution were meant to free the Indian farmer from constraints imposed by nature. Instead, large scale monocultures of exotic varieties generated a new ecological vulnerability by reducing genetic diversity and destabilizing soil and water systems.

Punjab was chosen to be India's bread-basket through the Green Revolution, with high response seeds, misleadingly called high yielding varieties (HYVs). The Green Revolution led to a shift from earlier rotations of cereals, oilseeds, and pulses to a paddy-wheat rotation with intensive inputs of irrigation and chemicals. The paddy-wheat rotation has created an ecological backlash with serious problems of waterlogging in canal-irrigated regions and groundwater mining in tubewell irrigated regions. Further, the HYVs have led to large-scale micronutrient deficiencies in soils, particularly iron in paddy cultivated areas and manganese in wheat cultivated areas.

Table 6.2 is an indicative summary of the costs and benefits of the Green Revolution in Punjab.

Table 6.2: The Green Revolution in Punjab
Indicative costs and benefits

Costs

1. Decline in pulses production from 370 to 150 thousand metric tons between 1965 and 1980.[12]
2. Decline in oilseeds production from 214 to 176 thousand metric tons between 1965 and 1980.[12]
3. Destruction of genetic diversity with introduction of rice and wheat monocultures.
4. 40 new insect pests and 12 new diseases in rice monocultures.[10]
5. Soils degraded by salinity, soil toxicity, micro-nutrient deficiency.
6. 2.6 lakh hectares waterlogged[9]
7. Punjab floods in 1988 linked to Bhakra dam. 65% of l2,000 villages submerged, 34 lakh people affected, 1,500 people killed. Loss to state, Rs1,000 crore.[14] 50,000 hectares of land destroyed through sand deposits exceeding 60cms in some places.[15]

Benefits

1. Increase in rice production from 292 to 3228 thousand metric tons between 1965 and 1980.[11]
2. Increase in wheat production from 1916 to 7694 thousand metric tons between 1965 and 1980.[11]

These problems were built into the ecology of the HYV's even though they were anticipated. The high water demands of these seeds necessitated high water inputs, and hence the hazards of desertification – through water logging in some regions and dessication and aridisation in others. The high nutrient demands caused micronutrient deficiencies on the one hand, but were also unsustainable because increased applications of chemical fertilizers were needed to maintain yields, thus increasing costs without increasing returns. The demand of the HYV seeds for intensive and uniform inputs of water and chemicals also made large-scale monocultures an imperative. And monocultures being highly vulnerable to pests and diseases, a new cost was created for pesticide applications. The ecological instability inherent in HYV seeds was thus translated into economic nonviability. The miracle seeds were not such a miracle after all. It is in the background of this ecological destruction caused by monocultures that the call for the diversification of Punjab agriculture was made by the Johl Committee in 1985.[16] A policy for diversification involves an increase in the genetic diversity in cropping systems. However, the Pepsico project and the associated New Seed Policy which was announced in September 1988, threaten to further erode the genetic diversity in agriculture by narrowing the crop base, and increasing its ecological vulnerabilities through the introduction of exotic varieties of fruit and vegetable monocultures.

An integral part of the Pepsico project is the introduction of new varieties of fruits and vegetable seeds to be developed at the agro research centre being set up at Ludhiana. The claimed objective is to develop improved varieties of potato, tomato, and selected crops using biotechnologies like clonal propagation and tissue culture.

However, the experience of the Green Revolution shows that 'improved seeds' is a contextual term, and genetic improvement for one objective can be a loss in terms of other parameters in another context. Thus, for Pepsi, 'improving' the potato implies making it more appropriate to its processing plant. The processing of potatoes and grain will be with imported machinery, which is supposed to handle about 30,000 tonnes of potatoes and 1,600 tonnes of grains per annum. The processing plant will determine the potato varieties to be planted, which will displace the native table varieties of potatoes. It is not that India does not grow tomatoes and potatoes. These are however, table varieties, to be consumed directly. There is no production of processed varieties; Pepsi will introduce these. From the point of view of the processing industry, the shift from table varieties to processed varieties is an improvement. For the consumer, it is a loss because the shift transforms food into raw material, to be consumed only through processing, not directly.

Potatoes for processing are being introduced in the name of 'diversification': – but given the experience of potato cultivation in the US from where Pepsico technology is being transferred, it will lead to genetic uniformity and high vulnerability. Today in the US only 12 varieties of the 2,000 species of potato are cultivated. 40% of all potato cultivation is of a single variety – the Russet Burbank. In 1970, only 28% of America's total potato acreage was planted with this variety. Acres and acres of the same kind of potato is ecologically very vulnerable. With Pepsico's biotechnology research centre, the potatoes planted will be genetically identical since they will be reproduced vegetatively from a single plant variety. It matters little that the company's research farms develop 50,000 potential new potato varieties every year because the imperatives of seed and food processing demand uniformity in cultivation, and it is cultivated diversity or uniformity which is linked to vulnerability to disease and pestilence.

The Pepsico project will encourage genetic uniformity because it integrates seeds with processing, the farm with the factory. It is precisely such a push that led to the spread of the Russet Burbank with the US. The McDonald corporation needed the Russet Burbank because of its size. For example 40% of all McDonald fries must be two to three inches long, another 40% must be over three inches; and the remaining 20% can be under two inches – and the Russet Burbank fits perfectly.[17] The economic forces of food processing push cultivation to a single crop yielding uniformity inspite of the known dangers of genetic uniformity, threatening the ecological stability of agriculture more than it has been in the past.

The introduction of uniformity is justified as a trade-off for raising yields of horticultural crops miraculously. Pepsi's promotion literature states that 'yields of horticultural produce in India are substantially lower than international standards'. The project proposal for Pepsi Food argues that 'in Mexico, Pepsi's subsidiary, Sabritas launched a seed programme that increase potato yields by 58% – from 19 to 30 tonnes per hectare in three years.'[18] In India, comparable yields have been achieved by farmers and agricultural scientists. As Usha Menon reported, potato yields of more than 40 tonnes per hectare have been realised during field trials in Jalandhar by the Central Potato Research Institute.[19] Yields averaging about 50-60 tonnes per hectare are also achieved by Gujarat farmers, who grow their potatoes on river beds in Banaskantha district. Just as in the first Green Revolution, the existence of indigenous high yielding varieties of rice was denied to justify the introduction of high response varieties. The Pepsi project denies the achievement of Indian farmers and scientists, to make Pepsico's role indispensable.

And just as the destruction of genetic diversity in the first

The Violence of the Green Revolution

Green Revolution cannot be ecologically or economically justified, the introduction of new ecological risks through projects like Pepsico are also unjustified.

Horticultural crops, especially when cultivated as monocultures, require high doses of pesticides. The genetic uniformity of processed varieties will thus increase the ecological hazards of pesticide use. The biotechnological revolution in agriculture is presented with the promise of new 'miracles' of fertilizers and pest-free crops. The dominant assumption of the liberalised seed policy is that it would ensure the supply of the 'best available' seeds from 'anywhere in the world' and this would give a fresh spurt to our agricultural production.

As the experience of the Green Revolution in Punjab amply demonstrates, seeds based on exotic strains and dependent on high energy and chemical inputs make for ecological instability and high vulnerability. The new seed policy of 1988, which has brought the import of seeds and seed-processing under the purview of Open General Licence (OGL) is closely associated with shifts in agricultural policy as symbolized by the Pepsico project. These shifts aggravate the vulnerabilities created by the Green Revolution while dispensing with some of the earlier safeguards. Green Revolution seeds like wheat and rice, were seeds of staple food crops, even though they contributed to genetic erosion and the decline of crops like, pulses and oil seeds. The focus of the new seed policy is on seeds of flowers, vegetables and fruits, not on food staples. While the seed policy mentions import of seeds for crops like pulses, oilseeds, coarse grains, vegetables and fruits, the international seed market is exclusively oriented towards the development of crop varieties which are widely traded on global markets. The new seed policy will thus imply the erosion of indigenous crops and crop varieties, while introducing heavier use of toxic agrochemicals.

Biotechnology will not reduce the use of farm chemicals but increase them since breeding for pesticide and herbicide resistance is the dominant focus of biotechnology research in agricultural crops. For the seed-chemical multinationals, this makes commercial sense especially in the short run, since it is cheaper to adapt the plant to the chemical than to adapt the chemical to the plant. The cost of developing a new crop variety rarely reaches US$2 million whereas the cost of a new herbicide exceeds US$40 million.[20] Herbicide and pesticide resistance will also increase the integration of seeds/chemicals and the control of MNC's in agriculture. A number of major agrochemical companies are developing plants with resistance to their brand of herbicides. Soyabeans have been made resistant to Ciba-Geigy's Atrazine herbicides, and this has increase annual sales of the herbicide by US$120 million. Research is also being done to develop crop plants resistant to other herbicides such as Dupont's 'Gist' and 'Glean' and Monsanto's 'Round-up' which are lethal to most herbaceous plants and thus cannot be applied directly to crops. The successful development and sale of crop plants resistant to brand name herbicides will result in further economic concentration of the agro-industry market increasing the market power of transnational companies.

For the Indian farmer this strategy for employing more toxic chemicals on pesticide and herbicide resistant varieties is suicidal, in a literal sense. In India, thousands of people die annually as a result of pesticide poisoning. In 1987, more than 60 farmers in India's prime cotton growing area of Prakasam district in Andhra Pradesh committed suicide by consuming pesticides because of debts incurred for pesticide purchase.[21] The introduction of hybrid cotton created pest problems. Pesticide resistance resulted in epidemics of white-fly and boll worm, for which the peasants used more toxic and expensive pesticides, incurring heavy debts and being driven to suicide. Even when pesticides and herbicides do not kill people, they kill people's sources of livelihood. The most extreme example of this destruction is that of bathua, an important green leafy vegetable, with a very high nutritive value and rich in vitamin A, which grows as an associate of wheat. However, with intensive chemical fertilizer use, bathua becomes a major competitor of wheat and has been declared a 'weed' that is killed with herbicides and weedicides. 40,000 children in India go blind each year for lack of vitamin A,[22] and herbicides contribute to this tragedy by destroying the freely available sources of vitamin A. Thousands of rural women who make their living by basket and mat making, with wild reeds and grasses, are also losing their livelihoods because the increased use of herbicides is killing the reeds and grasses. The introduction of herbicide-resistant crops will increase herbicide use and thus increase the damage to economically and ecologically useful plant species. Herbicide resistance also excludes the possibility of rotational and mixed-cropping, which are essential for a sustainable and ecologically balanced agriculture, since the other crops would be destroyed by the herbicide. US estimates now show a loss of US$4 billion per annum due to crop loss as a result of herbicide spraying. The destruction in India will be far greater because of higher plant diversity, and the prevalence of diverse occupations based on plants and biomass.

Strategies for genetic engineering for herbicide resistance which are destroying useful species of plants can also end up creating superweeds. There is an intimate relationship between weeds and crops, especially in the tropics where weedy and cultivated varieties have genetically interacted over centuries and hybridise freely to produce new varieties. Genes for herbicide tolerance, pest-resistance, stress-tolerance that genetic engineers are striving to introduce into crop plants may be transferred to neighboring weeds as a result of naturally occurring gene transfer.[23]

The outcome of the free import of genetically engineered seeds and crop varieties will lead to a drastic increase in the requirements for chemical herbicides for use on herbicide-resistant crops develope by agrochemical companies. The connsumption of agrochemicals to overcome the super-weeds associated with the Green Revolution has been ecologically and economically a disaster for peasants. The increased use of pesticides and herbicides with the introduction of crops engineered for herbicide tolerance will spell total doom.

The alternative strategy of genetic engineering of pest-resistant crops is not commercially desirable for agrichemical concerns in the short run. It is also not ecologically infallible in the long run, since genes for disease resistance can mutate, or they can be overcome by other environmental pressures, leaving the crop vulnerable. Introduced crops are anyway more prone to pest and disease attacks then native varieties, and they often introduce new pests and diseases in ecosystems.

The release of bio-engineered seeds needs to be viewed in the context of the historical experience with the Green Revolution varieties which brought with them new diseases and pests, even while it was claimed that new varieties were bred for resistance to pests. Since 1966, when new rice varieties were released in Punjab, 40 new insects and 12 new diseases have appeared. TN(I) the first semidwarf variety released in 1966, was susceptible to bacterial blight. In 1968, IRB, which was considered resistant to stem rot and brown spoot was released, but proved to be susceptible to both diseases. PR 106, PR 108, PR 109 were especially bred for disease and insect resistance. Since 1976, PR 106 has become susceptible to white-backed plant hopper, stem rot disease, rice leaf folder, hispa, stem border, and several other insect pests.[24] There is clearly no invulnerability in breeding for disease resistance. The more the technological claim to invulnerability, the greater is the ecological creation of vulnerability.

The liberalised seed policy has reduced the checks, controls, and safeguards for the release of new seeds while it opens the door for the introduction of new ecological hazards through the new varieties. The procedures laid down under the new policy for testing and trial of the imported seeds are weaker than those prevalent during the Green Revolution phase, when breeding technologies were less hazardous, and seeds used to be imported from the international agricultural research centres, not from MNC's and private breeders.

Introduction of diseases with introduced plant varieties is the norm, not the exception. For 10 years McDonald's has been trying to move the Russet Burbank to Europe, inspite of hundreds of European varieties being cultivated. When McDonald's tried to introduce the Russet variety in Holland in 1981, the potatoes had to sit in quarantine for eight months before they could be given trial plantings. But the potatoes proved vulnerable to the European potato virus, and was not accepted in Europe.[25] Such safeguards for rejection are being diluted in India. The usual norm for commercial release of seeds produced under alien agro-climatic conditions is three successive years of trial for occurrence of pests and diseases. The new seed policy has reduced this to a one season trial, which increases the risk of introduction of disease and pests. Further, all requirements for post-entry quarantine have been removed. Thus if crop diseases escape the quarantine check at entry, they can create havoc in the country's ecosystems.

Finally, in the case of genetically-engineered seeds and planting material, conventional quarantine methodologies are inadequate since the transgenic material itself is a source of ecological hazards.

Our risk-assessment frameworks are highly inadequate for the task of assessing the impact of the deliberate release of new plant variaties on ecosystems. Instead of developing methodologies for assessing risk, and strengthening regulations and structures for the protection of people's health and safety, the government has diluted quarantine procedures and removed import controls. Liberalisation has meant freedom for corporate giants to test, experiment and sell their products without constraint, without controls. This necessarily means destroying for citizens the right to freedom from hazards posed by the new technologies and products.

Seeds and Dependency

The biotechnology revolution, which spurs the shifts in India's seed policy, differs from the Green Revolution, in terms of corporate control and the control of bioregions. The biotechnology revolution is predominantly private in character. The Green Revolution was spearheaded by the international agricultural research centres like CIMMYT and IRRI organised by the Consultative Group on International Agricultural Research which is controlled by governments, private foundations, agribusiness corporations and multinational development banks. The private corporate interests such as agrochemical and agribusiness transnationals thus functioned through the programme set by public or quasi-public institutions, which they could influence and from whose agricultural strategies they stood to gain.

With the biotechnology revolution, the private corporate

multinational interest has become the spearheading sector of agricultural policy. The Pepsico Project and the new seed policy signals this new trend in which the technologies are not transferred from CIMMYT or IRRI to ICAR or PAU and on to the farm. This time, transnational corporate capital will go directly with the latest technology to the remotest farm. Private interests of profits will thus be the dominant driving force in the bio-revolution, increasing the control of multinationals, decreasing the role of governments and citizens of the Third World (Table 6.3).[28]

Table 6.3: Comparing the Revolutions

Green Revolution	Gene Revolution
Summary	
– Based in public sector	– Based in private sector
– Humanitarian intent	– Profit motive
– Centralised R & D	– Centralised R & D
– Focus of yield	– Focus on inputs/processing
– Relatively graduate	– Relatively immediate
– Emphasis on major cereals	– Affects all species
Objective	
– To feed the hungry and cool Third World political tension by increasing food yields with fertilizers and seeds	– To contribute to profit by increasing input and/or processor efficiencies
For whom	
– The poor	– The shareholder and management
By whom	
– CGIAR has 830 scientist working in 8 institutes reporting to US foundations	– In the USA alone, 1,127 scientists working for 30 agbiotech companies
– Industrialised countries	
– Quasi-UN bodies	
How	
– Plant breeding in wheat, maize, rice	– Genetic manipulation of all plants, all animals, microorganisms

(continued next page)

Green Revolution	Gene Revolution

Primary Targets

– Semi-dwarf capacity in – Response to fertilizers	– Herbicide tolerance – Natural substitution – Factory production

Investment

– $108 million for agricultural R&D through CGIAR (1988)	– Agbiotech R&D investment of $144 million in USA system (1988) by 30 companies

General Impact

– Substantial but gradual – 52.9% of Third World wheat and rice in HYV's (123 million hectares) – 500 million would not otherwise be fed	– Enormous, sometimes immediate – $20 billion in medicinal and flavour/fragrance crops at risk. – Multi-billion dollar beverage, confectionery, sugar and vegetable oils trade could be lost

Impact on Farmers

– Access to seeds and inputs uneven – Small farmers lose land to larger farmers – New varieties improve yield but increase risk – Reduced prices	– Increased production costs. – Loss of some crops to factory farms – Input/processing efficiencies increase farmer risk – Overproduction and materials diversification

Impact on Farms

– Soil erosion due to heavy use of crop chemicals – Genetic erosion due to replacement of traditional varieties – Species loss due to overplanting of traditional crops with maize, wheat or rice – Pressure on water resource due to irrigation	– Continuation and possible acceleration of Green Revolution effects plus – Release of potentially uncontrollable new organisms into the environment – Genetic erosion of animals and microorganisms – Biological warfare on economically important crops

(continued next page)

Green Revolution	Gene Revolution
Impact on consumption	
– Decline in use of high-value foods for poor people – Export of food out of region	– Emphasis on feeding the rich 'Yuppie' market. – Increased use of chemical and biological toxins
Economic Implications	
– Direct contribution of $10 billion per annum to Third World food production – Indirect contribution of $50-60 billion – Gene flow to US alone contribute to farms sales of $2 billion per annum for wheat, rice and maize.	– Contribution to seed production of $12.1 billion per annum by year 2000 – Contribution to agriculture of $50 billion per annum by year 2000 – Absorb benefit of gene flow from the Third World
Political Implications	
– National breeding programme curtailed – Third World agriculture westernised – Germplasm benefits usurped – Dependency	– CGIAR system subverted to corporate interests. – Genetic raw materials and technologies controlled by genetics supply industry through patents

Source: *Development Dialogue, 1988.*

The New Seed Policy seems to repeat the old mistakes of the Green Revolution of selling false miracles and threatens to render totally uneconomic the cultivation of staple food grains for local consumption by small farmers, thus threatening our food security as a nation. The dependence on import of seeds on the one hand and export of processed foods on the other has the very real danger of creating new forms of poverty and deprivation within the country, and making us totally dependent on a handful of multinational interests for the supply of inputs and the purchase of our agricultural commodities. The Pepsico project for the lab-to-farm-to-factory integration of seeds and agro-processing is an example of what the new liberalisation implies. As part of this integrated project, Pepsico will start a biotechnology base agro research centre for developing high-yielding disease resistant seeds of fruit and vegetable crops which the Pepsico plant will process. It took more than two years for Pepsico to get the clearance due to opposition by the public and by local industrialists. But with the New Seed Policy, doors have been opened for other multinationals.

The pharmaceutical giant, Sandoz India, has entered into an agreement with Northrup King of the US, subsidiary of its multinational parent company, and also with the Dutch vegetable king, Zaaduine. ITC is trying up with Pacific Seeds, a subsidiary of Continental Grains from Australia. The US seed giant Cargill Inc has tied up with the Gilland Company retaining controlling interest of the company. Two other US companies, Seedtec International and Dehlgien, have entered into agreements with Maharashtra Hybrid and Nath Seed Company respectively. Pioneer Hi-bred has started an Indian Subsidiary Pioneer Seed Company. Apart from these, Hindustan Lever is negotiating with a Belgium firm while Hoechst, Ciba-Geigy are reportedly moving in with other tie-ups.[29]

The Indian research stations and public sector seed producers like the National Seeds Corporation, which were until recently applauded for the Green Revolution, are now being reprimanded. And this new stance of the government towards the public sector is being used by it to legitimise the privatisation and transnationalisation of the seed industry. Naturally, the scientific community is alienated. At a seminar on the New Seed Policy at the National Institute of Science, Technology and Development (NISTAD) scientists of the Indian Agricultural Research Institute, the Council for Scientific & Industrial Research, the Central Food Technology Research Institute and the National Board for Plant

Genetic Resources responded critically to the New Seed
Policy.[30] These scientific bodies had not been consulted before
the Policy was announced. Indigenous research, and indigenous genetic resources have been sacrificed for corporate
research and the corporate supply of genetic material. Total
world retail sales in seeds per annum approximates US$13.6
billion, of which US$6 billion is proprietary (hybrid or pat-
ented seed). Analysts suggest that by the year 2000, the
world seed market will be US$28 billion, and US$12 billion
of this will be based on contributions from bio-technology.
With the opening up of the untapped Indian market, this
share will increase as will the experimental ground for trying
out new genetically engineered seeds with the ecological
risks they carry. As a recent issue of Development Dialogue
on the new bio-technologies points out, most of these agricultural inputs from genetic engineering are not 'here' yet
– they are 'arriving'. However, our new seed policy has
opened the door for that arrival, and has already allowed the
entry of the multinationals whose corporate strategies and
future profits hang on biotechnology.

The false miracle that seed companies are selling with
biotechnology and genetic engineering is the possibility of
liberating agriculture from chemicals and other ecological
risks. However, most of the seed multinationals are also
leading chemical companies. These include Ciba-Geigy, ICI,
Monsanto, Hoechst. The immediate strategy for these companies is to increase the use of pesticides and herbicides by
developing pesticide and herbicide tolerant varieties.

While deepening corporate control of agriculture, the
biotechnologies also expand the scope for this control. While
markets for agrochemical inputs and HYV seeds were restricted to regions with irrigation, the Bio-revolution will
permit the extension of commercial agriculture to all regions,
to rainfed lands and marginal soils. The impacts of the Bio-

revolution thus have the potential to encompass the entire rural populations of the Third World. Transnationals will thus gain total market control in a sector around which the life and livelihood of millions of farmers and peasants revolves. Given the private and proprietary character of the new technologies, the passage of the Plant Variety Protection Act and patenting of genetically modified life forms, the development of biotechnologies will bring tropical genes and tropical land under MNC control and will increasingly exclude Third World people from directly drawing sustenance from their land and the common heritage of living genetic wealth given by nature in abundant diversity in the tropics.

While the Green Revolution and Bio-revolution differ in scope and impact of control, they share the logic of commoditisation and demand-led growth in agriculture. The continuity is provided by the agri-business and agrochemical MNC's which controlled the Green Revolution indirectly but lead the Biotechnology directly and overtly. If the first line of products they sold to innocent nations and farmers is already known to be a failure, should not one be at least partially skeptical about the second line of products that multinational agribusiness is pushing through biotechnologies? If the package is technologically and financially beyond the access of the ordinary cultivator and people by its research and resource intensity, biotechnologies will breed new inequalities and new ecological hazards. In the Bio-revolution, as in the Green Revolution, 'improved seeds' will create a new dependence on global monopolies in the seed business.

Indigenous breeding through selection has given access to the best seeds to all, and the crop itself provides the seeds. Geertz's work on involution, and Richaria's work on conserving indigenous strains has established that by maintaining control over seeds, the peasant need not sacrifice in terms of yields. It is not the yields but the transnational control that 'improved' seeds improve. What is a drawback for the peasant is an advantage for the seed corporation. The hybrid seed must be bought each year from the seed merchants. The genetically engineered 'seed' of the Bio-revolution will deepen this dependence of peasants on MNC's. The lab, not nature, will become the sole source of seeds of the biotechnologies, and with labs shifting from universities to the corporate sector, from the public to the private domain, only those who can pay, will have a right to seeds. As the integrations with nature as the primary genetic source are broken down, new integrations appear within MNC's. Biotechnologies, seeds, and agrichemicals are merging together. Corporate strategies are using agricultural bio-chemistry and biotechnology to get a new monopoly over plant breeding. As mentioned earlier, genetically engineered plant varieties are being developed by seed companies to be compatible with the proprietary plant protection chemicals manufactured by another subsidiary of the same company. Monsanto has genetically engineered a tomato variety to be compatible with one of its herbicides, 'Round-Up'. Calgene has recently obtained a patent for a modified DNA chain which enhances the herbicide resistance of plants. With a single patent, they now have exclusive rights to charge others in utilising that property whether in tomatoes, tobacco, soyabeans, cotton, etc. Biotechnology could thus make possible the full integration of biological and chemical product lines, and thus a complete control on agriculture and on genetic resources.[31]

The Pepsico project and the New Seed Policy is supposed to bring the latest biotechnologies to India. But it is in the nature of these technologies that they will be capital intensive and research intensive. They will be physically located in India, but controlled by multinationals of industrialised countries like Pepsico of USA.

Recent pressures for liberalising the patent laws of India are closely related to getting monopoly control, not just over plants, but also over plant traits.[32] The US has set a precedence in granting utility patents which allow claims for only part of a plant, or the genes expressing certain traits, like black flowers, tolerance to salt water, the ability to produce nitrogen etc.This gives the patent holder the right to exclude or collect a royalty from others reproducing any plant or selling seeds carrying the patented trait. A plant breeder wanting to utilize those genes in another variety or species would require a license. Utility patents would also make the practice of planting second generation seeds from a protected variety illegal. It is expected that utility patents will double the current seed bills of farmers. A utility patent has been granted for 17 years to a seed company which had utilized genetic engineering to introduce the protein trytoplah into corn, thus increasing its nutritional value.[33]

Patents and other intellectual property rights are the remaining hurdles to be crossed for large scale distribution of biotechnology seeds by transnational corporations. For instance, one of the Clauses of the New Seed Policy directs all companies importing seeds to make a small quantity available to the gene bank of the government controlled National Bureau of Plant Genetic Resources (NBPGR). The corporate giants are, of course, unwilling to accept that clause and want its removal. As Jan Nefkins, General Manager of Cargill South-east Asia Limited, pointed out: 'No Company would be willing to part with what they took years and spent millions of dollars developing. It's a question of intellectual property rights.'[34]

The fight over patents and proprietary rights has become an essential element of global politics in the biotechnology era. On the one hand, the US is trying to globally introduce its patent protection system, which is heavily biased in favour of the industrially developed countries. Under the 1984 amendment to the trade-act, the US government considers the lack of patent protection to transnationals as unfair trading practice, and is using trade as a weapon in the battle over patents. On the other hand scientists, industrialists and public interest groups in countries like India are demanding a use of India's patent laws for protection of the national and public interest. The National Working Group on Patent Laws has been critical of attempts to erode national sovereignty through changes in patent laws. New economic conflicts between industrialized countries and the Third World, and between private corporate interest and the public interest are emerging with the age of bio-engineering. Bio-technologies are opening up new areas for corporate profit, and even while the promise is of unprecedented prosperity, like the Green Revolution, it could be a prosperity with a very high price, especially for the poor.

Seeds of Insecurity, Seeds of Violence

Globally and nationally, food grain production has been dramatically reduced due to ecological instabilities. These include drought, induced both by climatic change associated with the Greenhouse effect, and desertification through inappropriate land and water use. The momentum of grain production between 1950 and 1984, when world output grew from 624 million tonnes to 1,645 million tonnes has waned in the late 1980s and may continue to dip in the nineties. The world carry-over stocks at the beginning of the harvest season in 1989 will not be more than 243 million tonnes, just enough to feed the world's population for 54 days against 459 million tonnes to meet 101 days' consumption needs in 1987.[35] In India food grain stocks were at rock bottom at the end of 1988. The country had food stocks of barely 7.7 million tonnes on 1 November, 1988 i.e. 2.2 million

Table 6.4: United States Grain Production, Consumption and Exportable Surplus by Crop Year, 1984-88

Year	Production	Consumption Surplus from Current Crop	Exportable
		(million metric tons)	
1984	313	197	116
1985	345	201	144
1986	314	216	97
1987	277	211	66
1988	190	202	·2

Note: ¹ Does not include carry-over stocks.

Sources: *US Department of Agriculture, Economic Research Service, World Grain Harvested Area, Production, and Yield 1950-87 (unpublished printout) (Washington, DC , 1988); USDA, Foreign Agricultural Service, World Grain Situation and Outlook, August1988.*

tonnes of rice and 5.5 million tonnes of wheat. This was against 15.7 million tonnes of food grains – 5.4 million tonnes of rice and 10.3 million tonnes of wheat – held by the country on 1 November, 1987.[36]

The public distribution system, which supplies subsidised food grain, requires 7 lakh tonnes of rice and six lakh tonnes of wheat every month. Besides this, grain needs are distributed through the National Rural Employment Programme (NREP) and the Rural Landless Employment Guarantee Programme (RLEGP), which aim at providing for the survival needs of the poorest. A large part of food grain for the public distribution system comes from Punjab. Production of surplus in Punjab for the food needs of the entire country was the objective of the first Green Revolution. An indication of Punjab's contribution to India's national stock is that 85.7% of the paddy and 57.3% of the wheat produced in 1985-86 was procured by various agencies in the state.

What will be the impact on the availability of staple foods as more and more land is diverted to fruits and vegetables for export, at a time when food scarcity is already a reality, both nationally and globally? Will potato chips for export feed the hungry in India who will be further deprived of grain through the public distribution system? And with the pressure from the World Bank and the IMF to reduce food subsidies, how will the food entitlements of the economically and politically weakest groups be protected?

During the visit of the Director of the IMF, Mr Camdessus, in October 1988, the Union Minister for Finance had indicated that India would be reducing subsidies for food distribution. The Finance Ministry's willingness to fall in line with World Bank/IMF policy was resisted by the Food Ministry which saw the issue of food subsidies as essential for maintaining the public distribution system.[37]

The experience of countries in Latin America and Africa where food riots have broken out following IMF/World Bank conditionalities for the removal of subsidies can help us anticipate the violent consequences of the new agricultural policy symbolised by the Pepsico project, which introduces new subsidies for processed food exports, while removing subsidies for the domestic distribution of staple foods.

Higher farm incomes are a primary promise of the Pepsico project. Promotion literature from Pepsico and the new Ministry for Agroprocessing announces that farm incomes which are around Rs7,000 per hectare from cereal cultivation may jump to Rs15,000–Rs20,000 per hectare from fruit and vegetable cultivation.[38]

The experience of the first Green Revolution should teach us to treat such economic miracles with caution. The profitability in the cultivation of wheat and rice of the early period of the Green Revolution did not last too long, since increased inputs were needed to maintain yields of the Green Revolution varieties. According to a recent study, the net returns from wheat cultivation per hectare at 1970-71 prices declined from Rs328 in 1971-72 to Rs54 in 1981-82. As a result, the profitability of large farmers declined and the deficits of the poor peasants have risen.[39]

The miraculous jumps in farm incomes as promised by the Pepsico project and the new agricultural policy will be similarly short-lived even for the small group of farmers that benefits from it. Farmers who started cultivating hybrid tomatoes in the Western region of India initially had incomes as high as Rs30,000 per hectare because hybrid tomatoes give yields of about 40 tonnes per hectare as compared to 22 tonnes per hectare produced by open pollinated varieties. However, hybrids are also more vulnerable to pests and disease, and in a few years incomes from hybrid tomato cultivation had dropped from Rs30,000 to a few hundred rupees.

Cash crops, especially for export, are subject not just to ecological risks, but to financial risks as well, because cash crops for export do not produce much cash over time. The growth of export-oriented cash crop agriculture is a primary reason for Africa's food crisis. As Lloyd Timber lake states in the context of Africa's food crisis, 'the main drawback to cash crops is that over the past decade they have produced less and less cash.'[40] First, cash crops are encouraged over food production by an export oriented agriculture policy. As the area under commodities for exports grows, prices fall and returns decline instead of increasing. As a catalyst, the Pepsico Project for a new export-oriented agriculture policy, will put India on the path of debt, dispossession and agricultural decline that the export-led agricultural strategies have

The 'intended outcome' of course! (margin annotation)

created in Africa and Latin America.

As Clairmonte and Cavanagh observed,

> 'The outcome, like tragedy, is ineluctable; Third World
> countries are literally being driven to market fatter and
> fatter volumes of commodities at lower and lower prices on
> the global market in return for higher priced goods and
> service imports.'[41]

Cash-crop exports have been tried elsewhere and are a proven way to get trapped in food scarcity and spiraling debt burdens. Africa's food crisis and hunger and famine are linked directly to the underdevelopment of Africa's food production by cash-crops leading to a decline in food production. Scarce resources have been diverted to each crop, undermining the cultivation of food and causing major ecological instability. As recently as 1970, Africa was producing enough food to feed itself. By 1984, 140 million Africans out of a total of 531 million were fed with grain from abroad, because by the end of the 1970s, the economies of many African nations were tied to cash-crop production. Dependence on single crop commodities for export is in large measure at the root of Africa's ecological, economic and human crisis.

Pepsico's entry will generate a similar crisis for India as the logic of diverting land from staple food to processed food exports will unfold. It took Africa less than a decade to go from food sufficiency to scarcity. It could take India less than that if the Pepsico Project sets precedence, and other multinational agribusinesses move in to use India's farmland for export commodities. Food companies that are already looking at India for investment include Kellogs, Campbell, Heinz, Del monte, Nestle, Nissan, Swedish Match AB among many others.[42] The impact of integrated agribusiness projects on food entitlements of people in the Third World has been rehearsed too often for us to close our eyes to it. The issue is not the entry of a soft drink multinational, but of the control by a multinational agribusiness of India's food growing land, largely for export commodities. Is India ready to forego its carefully built strategy for self-reliance in food?

The experience of the first Green Revolution shows that price incentives and subsidies for the production of the new varieties of wheat and rice were simultaneously disincentives and penalties for the production of coarse grains, pulses and oilseeds. The scarcity of oil-seeds has become so severe that the government has had to create a special 'oilseed technology mission' to increase cultivation and production. The introduction of incentives and subsidies for production of fruits and vegetables as raw material for the export-oriented-agro-processing industry will similarly act as a disincentive for the production of staple food grains, aggravating the present scarcity and diseconomies.

Employment generation is a much exaggerated economic benefit associated with the Pepsico project. The project itself will employ only 489 people but employment generation through indirect job creation is put at 25,000 in Punjab agriculture and another 25,000 in the rest of the country. The official figures fail to take into account the large scale unemployment that will be created by the displacement of small and marginal farmers and among the self-employed.

Industrialisation of agriculture is a proven way towards aggravating rural unemployment rather than ameliorating it. Comparative research on 22 rice growing areas has shown that industrialisation displaces labour, while alternative strategies exist which increase productivity through increase of labour inputs.[43]

Integrated industrialisation of agriculture, from production to processing will also create large-scale labour displacement and loss of skill in the informal sector where most food processing currently takes place. The Pepsico advertisers ridicule Indian foods as being ethnic – 'made up of traditional mango pulp, pickles and chutney'. These products have suited our needs; our climate and our skills. Joining the global monoculture of potato chips will reduce our richness in diets, and destroy our food diversity. It will also lead to large scale de-industrialisation and 'de-skilling' in the traditional food processing industry, in which every home and every community in the country participates productively.

The irony of the capital intensive, high input agricultural strategy that was initiated with the Green Revolution, and is being carried to the next stage with the biotechnology and food processing revolution, is that it generates violence and distress not only for those for whom it fails but even where it succeeds. It creates social, political and economic crises by generating scarcity on the one hand and generating surpluses on the other. The crisis of scarcity and the crisis of surpluses are two aspects of the same crisis generated by a non-sustainable resource and capital intensive agriculture. Small farmers are victims of both aspects of the crisis, in the North and in the South, in industrialised countries as well as in largely agrarian societies.

According to Wendell Berry,

> 'For nothing at present is more destructive of farms and farmers than bumper crops. High production keeps production costs high and the market prices low – a bonanza (so far) for the industrial suppliers of "purchased inputs" to farmers and (so far) to the banks from which desperate farmers must borrow (at usurpious interest) the increasing amounts of money necessary to bridge the gap between

the depressed farm economy and an inflated industrial economy. But high production is death to farmers. Though most interested parties seem still addicted to the perception of American agricultural productivity as a "miracle", even news articles now occasionally glimpse the fact that, for farmers, a good year in corn is a bad year in dollars.'[44]

A *Newsweek* article entitled 'Down and Out on the Farm' reported how farmers were being dispossessed by the 'success' of increasing production.[45] In Ontario, Canada, the government has started a $6 million relief programme for farmers who have been left redundant in a capital and resource intensive industrialised farming. The programme is called 'Farmers in Transition', and includes a help line (1 -800-265-1511) for free, around-the-clock-telephone assistance to depressed and distressed farmers.[46] Suicide rates and violence have shot up in rural areas of North America as part of the crisis.

As Mark Ritchie and Kevin Ristau reported,

'the agricultural crisis in affluent America is sowing seeds of violence, and 'many bitter and desperate rural people, faced with losing everything they've worked for, may become involved in one of the extremist organizations that are increasingly active throughout the countryside.'[47]

Agricultural surpluses of selected commodities created artificially through high capital, energy and resource inputs seem to generate invisible scarcity, rising distress and conflict. The violence of Punjab seems to be the pattern associated with capital and resource intensive agriculture everywhere.

The first technological fix of the Green Revolution in Punjab brought violence instead of peace, dependence instead of self-reliance and autonomy. The second technological fix heralded by Pepsico could deepen the trends on violence and dependence because it will be more excluding, more centralising and more globally integrated into international markets of farm inputs and farm commodities. In the final analysis, the deepening violence and militarisation of society is linked with a loss of autonomy and loss of control of people over their lives. It arises from the violation of the integrity of people as they are turned into part of the assembly line of corporate food production or thrown out as waste. The path to peace cannot come from deepening the centralising thrust that is at the root of present discontent. Restoring a decentered and ecological system of food production is the only strategy for creating a lasting peace.

An agriculture integrated from farm - to - factory by the same corporate interests creates new risks and new vulnerabilities for Indian peasants, and consumers. The corresponding disintegration of ecological and political processes will aggravate the discontent that already exists. Pepsi and the new agricultural policy cannot become a programme for peace because it establishes more centralised control over farmers' lives, and introduces new instability in agricultural systems at the ecological, economic and political levels. It cannot fulfill 'the promise of a hundred years of spring'.

References 6

1.   R H Richaria, paper presented at seminar on 'Crisis in Modern Science', Penang, November 1986.

2.   A R Yegna Iyengar, *Field Crops of India*, Bangalore: BAPPCO, 1944, p30.

3.   M S Gill, 'Success in the Indian Punjab', in J G Hawkes, *Conservation and Agriculture*, London: Duckworth, 1978, p193.

4. D S Kang, 'Environmental Problems of the Green Revolution with a focus on Punjab, India', in Richard Barrett, (ed), *International Dimensions of the Environmental Crisis*, Boulder: Westview Press, 1982, p204.

5. Gunvant Desai, 'Fertilizers in India's Agricultural Development', in C H Shah, *Agricultural Development of India*, Orient Longman, 1979, p390.

6. T B Bayliss Smith, 'The Green Revolution at micro scale', *Understanding Green Revolutions*, Cambridge University Press, p984.

7. Prem Shankar Jha, 'The Pepsi Project', *Times of India*, 1 September, 1986. 'Punjab: Programme for Peace', *Hindustan Times*, 11 December, l986.

8. Pramod Kumar, *et al*, *Punjab crisis: Context and Trends*, Centre for Research in Rural and Industrial Development, Chandigarh, 1984.

9. Punjab Agricultural University, Department of Soils.

10. G S Sidhu, *The Green Revolution and Rice diseases in Punjab'*, mimeo.

11. 'Punjab Floods were Manmade', *Economic Times*, 4 October, 1988.

12. S S Johl, 'Diversification of Punjab Agriculture', Government of Punjab, 1985.

13. Flood Reduce Cultivable area, *Economic Times*, 26 January, 1989.

14. A Bhattacharjee, 'New Seed Policy: Whose interest would it serve', *Economic and Political Weekly*, 8 October, 1988, p2089.

15. Floods Reduce Cultivable area, *Economic Times*, 26 January, 1989.

16. S S Johl, *Diversification of Punjab Agriculture*, Government of Punjab, 1985.

17. A Bhattacharjee, 'New Seed Policy: Whose interest would it serve', *Economic and Political Weekly*, 8 October, 1988, p2089.

18. Jack Doyle, *Altered Harvest*, New York: Viking, 1985, p205.

19. Usha Menon, *Anything for a dollar: A Close Look at the Pepsi Deal*, Delhi Science Forum, 1989.

20. Cary Fowler, *et al*, 'Laws of Life', *Development Dialogue*, Uppsala: Dag Hammarskjold Foundation, 1988.

21. Vanaja Ramprasad, *Hidden Hunger*, Research Foundation for Science and Ecology, 1988.

22.   Mira Shiva, personal communication.

23.   Peter Wheale and Ruth McNally, *Genetic-Engineering: Catastrophe or Utopia*, UK: Harvester, 1988, p172.

24.   G S Sidhu, 'Green Revolution and Rice diseases in Punjab', mimeo.

25.   Jack Doyle, *Altered Harvest, op cit*, p207.

26.   John W Rosenblum, (ed), *Agriculture in the 21st Century*, New York: Wiley – Interscience, 1983.

27.   Martin Kenny, *Biotechnology: University Industry Linkages*, Yale, 1986.

28.   Cary Fowler, *op cit*.

29.   'Seeds: A hard row to hoe', *India Today*, 15 February, 1989.

30.   'Scientists Seeth at Seed Policy,' *Economic Times*, 1 December, 1989.

31.   *Patents and plant genetic Resources*, Foundation for Economic Trends, Washington, mimeo, 1986.

32.   Proceedings of National Seminar on Plant Laws, 22 November, 1989, organised by National Working Group on Patents Laws, New Delhi.

33.   *Patents & Plant Genetic Resources*, Foundation for Economic Trends, Washington, mimeo, 1988.

34.   'Seeds: A hard row to hoe', *India Today*, 15 February, 1989.

35.   Lester R Brown, *The Changing World Food Prospect: The Nineties and Beyond*, World Watch Paper 85, World Watch Institute, October 1988.

36.   'Food Grains Stock at Rock Bottom', *Economic Times*, 15 December, 1988.

37.   *Fertiliser, food subsidies may be cut*, Hindu, 12 October, 1988.

38.   *Triumph of Commitment*, Pepsi Foods Ltd.

39.   S S Gill, 'Contradictions of Punjab Model of Growth and the Search for an Alternative', *Economic and Political Weekly*, 15 October, 1988.

      S S Gill, 'The Price of Prosperity Problems of Punjab's Agriculture', *Times of India*, January 1989.

40.   Lloyd Timberlake, *Africa in Crisis*, London: Earthscan, 1985.

41.   Clairmonte and Cavanagh, 'Third World Debt: The Approaching Holocaust', *Economic and Political Weekly*, 2 August, 1986.

42. 'Foreign firms queue up for food processing tie-ups', *Economic Times*, 2 April,1989.

43. Tim Bayliss Smith, *op cit*, p169.

44. W Berry, 'Whose Head is the Farmer Using?' in W Jackson, W Berry, Bruce Coleman, *Meeting the Expectations of the Land*, San Francisco: North Point Press, 1984.

45. *Newsweek*, 12 April, 1982.

46. *Farmers in Transition*, Ministry of Agriculture and Food, Ontario 1986.

47. Mark Ritchie and Kevin Ristau, *Crisis by Design: A Brief Review of US Farm Policy*, League of Rural Voters Education Project, Minneapolis, 1987.

7

THE SEED AND THE SPINNING WHEEL:
The Political Ecology of Technological Change

IN the dominant paradigm, technology is seen as being above society both in its structure and its evolution, in its offering technological fixes, and in its technological determinism. It is seen as a source of solution to problems that lie in society, and is rarely perceived as a source of new social problems. Its course is viewed as being self-determined. In periods of rapid technological transformation, it is assumed that society and people must adjust to that change, instead of technological change adjusting to the social values of equity, sustainability and participation.

There is, however, another perspective which treats technological change as a process that is shaped by and serves the priorities of whomever controls it. In this perspective, a narrow social base of technological choice excludes human concerns and public participation. The interests of that base are protected in the name of sustaining an inherently progressive and socially neutral technology. On the other hand, a broader social base protects human rights and the environment by widening the circle of control beyond the current small group.

The emergence of the new biotechnologies brings out these two tendencies dramatically. The technocratic approach to bio-technology portrays the evolution of the technology as self-determined and views social sacrifice as a necessity. Human rights including people's right to livelihoods therefore must be sacrificed for property rights that give protection to the innovation processes. Ironically, a process based on sacrifice of human rights continues to be projected as automatically leading to human well-being.

The sacrifice of people's rights to create new property rights is not new. It has been part of the hidden history of the rise of capitalism and its technological structures. The laws of private property which arose during the fifteenth and sixteenth centuries simultaneously eroded people's common rights to the use of forests and pastures while creating the social conditions for capital accumulation through industrialisation. The new laws of private property were aimed at protecting individual rights to property as a commodity, while destroying collective rights to commons as a basis of sustenance. The Latin root of private property, privare, means 'to deprive'. The shift from human rights to private property rights is therefore a general social and political precondition for exclusivist technologies to take root in society. The scene for such a shift is now being set to allow the emergence of a biotechnological era of corporate and industrial growth.

In the narrow view, science and technology are conventionally accepted as what scientists and technologies produce, and development is accepted as what science and technology produce. Scientists and technologists are in turn taken to be that sociological category formally trained in Western science and technology, either in institutions or organisations in the West, or in Asian institutions mimicing the paradigms of the West. These tautalogical definitions are unproblematic if one leaves out people, especially poor people, if one ignores cultural diversity and distinct civilisational histories of our planet which has created diverse and distinctive cultures. Development in this view is taken as synonymous with the introduction of Western Science and Technology in non-Western contexts. The magical identity is development = modernisation = Westernisation.

In a wider context where, science is viewed as 'ways of knowing' and technology as 'ways of doing' all societies, in all their diversity, have had science and technology systems on which their distinct and diverse development has been based. Technologies or systems of technologies bridge the gap between nature's resources and human needs. Systems of knowledge and culture provide the framework for the perception of definition of science and technology – science and technology are no longer viewed as uniquely Western but as a plurality associated with all cultures and civilisations. And a particular science and technology do not automatically translate into development everywhere. Ecologically and economically inappropriate science and technology can become causes of underdevelopment, not solutions to underdevelopment. Ecological inappropriateness is a mismatch between the ecological processes of nature which renew life support systems and the resource demands and impacts of technological processes. Technological processes can lead to higher withdrawals of natural resources or higher additions of pollutants than ecological limits allow. In such cases they contribute to underdevelopment through destruction of ecosystems.

Economic inappropriateness is the mismatch between

the needs of society and the requirements of a technological system. Technological processes create demands for raw materials and markets, and control over both raw materials and markets becomes an essential part of the politics of technological change.

The lack of the theoretical cognition of the two ends of technological processes, its beginning in natural resources and its end in basic human needs, has created the current paradigm for economic and technological development which demands increasing withdrawals of natural resources and generates increasing addition of pollutants while marginalising and dispossessing increasing numbers from the productive process. These characteristics of contemporary scientific industrial development are the primary causes for the contemporary ecological, political and economic crises. The combination of ecologically disruptive scientific and technological modes and the absence of criteria for evaluating scientific and technological systems in terms of resource use efficiency and capability for basic needs satisfaction has created conditions where society is increasingly propelled towards ecological and economic instability and has no rational and organised response to arrest and curtail these destructive tendencies.

The introduction of ecological and economically inappropriate science and technology leads to underdevelopment instead of development. Modernisation based on resource hungry processes materially deprives communities which use those resources for survival, either directly, or through their ecological function. Growth under these conditions does not ensure development for all. It creates underdevelopment for these affected negatively by resource diversion or destruction. Conflicting demands on resources thus lead to economic polarisation through growth. The growing extent of the people's ecology movements is a symptom of this polarisation and a reminder that natural resources play a vital role in the survival of the people. Their diversion to other uses or their destruction through other uses is therefore creating increasing impoverishment and an increasing threat to survival.

Underdevelopment is commonly projected as a state created by the absence of modern western science and technology systems. However, poverty and underdevelopment are, more often than not, conditions created by the externalised and invisible costs of resource intensive and resource destructive technological processes which support the livelihood of millions.

The experience of all industrial revolutions illustrates how poverty and underdevelopment is created as a integral part of the whole process of contemporary growth and development in which gains accrue to one section of the society or nation and the costs, economic or ecological are borne by the rest.

The first industrialisation was based on the mechanisation of work, with focus on the textile industry. The second industrialisation was based on the chemicalisation of processes in agriculture and other sectors, and the emerging third industrialisation is based on the engineering of life processes.

We can draw some lessons from history about how technological change initiated by a special interest brings development to that interest group while creating underdevelopment for others.

Colonisation and the Spinning Wheel

The mechanisation of textile manufacture was the leading technological transformation of the first industrial revolution. By the time that technological innovations made full impact on the British textile industry in the early 19th century, England had gained full political control over resources and its markets of its colonies; including India. India until then had been a leading producer and exporter of textiles in the world market. The industrialisation of England was based in part on the deindustrialisation of India. The development of England was based on India's underdevelopment. It is no co-incidence that India's independence movement was based in large measure on seeking liberation from the control on resources and people of Third World that were part of the process of Europe's industrialisation. Two symbols of India's independence struggle were the 'Champaran Satyagraha' and the 'Charkha'. The Champaran Satyagraha was a peaceful revolt against the forced cultivation of indigo as a dye for the British textile industry. The 'Charkha' or spinning wheel was the technological alternative that created self-reliance instead of dependence, and generated livelihoods instead of destroying them.

While the rapid technological innovations in the British textiles industry were made possible only through the prior control over the resources and the market, the stagnation and decay of this industry in India was a result of the loss of political control first over the market and later over the raw material. The destruction of India's textile industry necessitated the destruction of the skills and autonomy of India's weavers. Often this destruction was extremely violent. For instance, the thumbs of the best Bengal weavers were cut off to cut off market competition when Indian hand woven textiles continued to do better than the British mill products.[1] The impact of the violence manipulation and control of the

English merchants on the Indian weavers started when the East India Company became a territorial power by defeating Nawab Sirajuddaula in the battle of Palassi in 1757. Before that the Indian weavers were independent producers and had control over their produce. The East India company replaced the indigenous merchants by a 'body of paid servants receiving instructions from them with coercive authority over weavers that none had before. They had virtual monopoly of the market and had effectively exercised control over raw materials and began to extend this control over the weavers' tools. Under the company weavers had virtually become wage workers on terms and conditions over which they had no control.'[2]

In the context of such erosion of the control on resources and the market, the traditional weavers of India were displaced. There was an exodus out of the weaving trade. New textile technology was imported into India from England in the mid-19th century by the cotton traders of India who were involved in export of cotton to England. This new group of powerful merchants turned mill owners competed with the handloom weavers for the common market and the raw material base. The establishment of textile mills in Lancashire and later in India deprived the Indian weaver both of the market and the raw material. When the American cotton supply to the English textile industry was disturbed by the American civil war, the famous cotton famine of 1860s broke out and the English instantly reacted by grabbing the cotton in India. The cotton famine was transferred to India.

A government survey of 1864 gives the following picture of the production and supply of clothing:

'It is evident that the whole population must be far nearer a state of pristine nudity than ever before. Every poor person stints himself to an inconceivable degree in his

clothing and every purpose to which cotton is applied. He wears his turban and breach cloth to rags, dispenses with his body clothing and denies himself of his annual renewal of his scanty suit.' [3]

There was also a devastating impact of the new textile mills opened in India on the handloom weavers.

'The growth of the industry began to impinge on the handloom industry......This incursion of mills into areas hitherto considered the special reserves of the handloom industry had a many sided effect....and led to unprecedented worsening of the conditions of the handloom weavers..... Actual unemployment was seen as in the statistics of idle handlooms; this was estimated at 13% in 1940 by the fact finding committee (of Handlooms and mills).' [4]

Gandhi's critique of the industrialisation of India on the western model was based on his perception of the poverty, dispossession and destruction of livelihoods which resulted from it.

'Why must India become industrial in the western sense?', Gandhi has asked 'what is good for one nation situated in one condition is not necessarily good for another differently situated. One man's food is often another man's poison.... Mechanisation is good when hands are too few for the work intended to be accomplished. It is an evil where there are more hands than required for the work as is the case in India.' [5]

It was to regenerate livelihoods in India that Gandhi thought of the spinning wheel as a symbol of liberation and a tool for development. Power driven mills were the model of development in that period of early industrialisation. However, the hunger of mills for raw-material and markets was the reason for a new poverty, created by the destruction of livelihoods either by diverting land and biomass from local subsistence to the factory, or by displacing local production through the market.

Gandhi had said that 'anything that millions can do together, becomes charged with unique power'. The spinning wheel had become a symbol of such power. 'The wheel as such is lifeless, but when I invest it with symbolism, it becomes a living thing for me.'[6]

When Gandhi described the Charkha in 1908, in Hind-Swaraj as a panacea for the growing pauperism of India, he had never seen a spinning wheel. Even in 1915, when he returned to India from South Africa, he had not actually seen a spinning wheel. But he saw an essential element of freedom from colonialism in discarding the use of mill woven cloth. He set up handlooms in the Satyagraha Ashram at Sabarmati, but could not find a spinning wheel or a spinner, who were normally women. In 1917, Gandhi's disciple Gangabehn Majumdar started a search for the spinning wheel, and found one in Vijapur in the Baroda State. Quite a few people there had spinning wheels in their homes, but had long since consigned them to the lofts as useless lumbers. They now pulled them out, and soon Vijapur Khadi gained a name for itself. And Khadi and the spinning wheel rapidly become the symbol for India's independence movement.

The spinning wheel symbolised a technology that conserves resources, people's livelihoods and people's control over their livelihoods. In contrast to the imperialism of the British textile industry, the 'Charkha' was decentred and labour generating, not labour displacing. It needed people's hands and minds, instead of treating them as surplus, or as mere inputs into an industrial process. This critical mixture of decentralisation, livelihood generation, resource conservation and strengthening of self-reliance were essential to undo the waste of centralisation, livelihood destruction, resource depletion and creation of economic and political dependence that had been engendered by the industrialisation associated with colonialism.

Gandhi's spinning wheel is a challenge to notions of progress and obsolescence that arise from absolutism and false universalism in concepts of science and technology development. Obsolescence and waste are social constructs that have both a political and ecological component. Politically, the notion of obsolescence gets rid of people's control over their lives and livelihoods by defining productive work as unproductive and removing people's control over production in the name of progress. It would rather waste hands than waste time. Ecologically, too obsolescence destroys the regenerative capacity of nature by substituting manufactured uniformity in place of nature's diversity. This induced dispensability of poorer people on the one hand and diversity on the other constitutes of the political ecology of technological development guided by narrow and reductionist notions of productivity. Parochial notions of productivity, perceived as universal, rob people of control over their means of reproducing life and rob nature of her capacity to regenerate diversity.

Ecological erosion and destruction of livelihoods are linked to one another. Displacement of diversity and displacement of people's sources of sustenance both arise from a view of development and growth based on uniformity created through centralised control. In this process of control, reductionist science and technology act as handmaidens for economically powerful interests. The struggle between the factory and the spinning wheel continues as new technologies emerge.

The Colonisation of the Seed

The changes that took place in the textile industry during colonialism were replayed in agriculture after India's independence through the Green Revolution. Whether it was the chemicalisation of agriculture through the Green Revolution or its transformation through the new biotechnologies, the seed is at the center of all recent changes in agricultural production.

All technological transformation of biodiversity is justified in the language of 'improvement' and increase of 'economic value'. However, 'improvement' and 'value' are not neutral terms. They are contextual and value laden. What is improvement in one context is often regression in another. What is value added from one perspective is value lost from another.

The 'improvement' of the seed is not a neutral economic process. It is, more importantly a political process that shifts control over biological diversity from local peasants transnasional corporations and changes biological systems from complete systems reproducing themselves into raw material. It therefore changes the role of the agricultural producer and the role of ecological processes. The new biotechnologies follow the line of the path of hybridisation in changing the location of power as associated with the seed.

As Kloppenburg has stated,

'It decouples seed as 'seed' from seed as 'grain' and thereby facilitates the transformation of seed from a use-value to an exchange value.'[7]

Agricultural research is primarily a means of eliminating

barriers to the penetration of agriculture by capital. The most important barrier is the nature of the seed, which reproduces itself and multiplies. The seed thus possesses a dual character that links both ends of the process of crop production: It is both means of production and, as grain, the product. In planting each years crop the farmers also reproduce a necessary part of their means of production. The seed thus presents capital with a simple biological obstacle: Given appropriate conditions the seed will reproduce itself manifold.

The seed has therefore to be transformed materially if a market for seed has to be created.

Modern plant breeding is primarily an attempt to remove this biological obstacle to the market in seed. Seed reproducing itself stays free, a common resource and under the farmers control. Corporate seed has a cost and is under the control of the corporate sector or under the control of agricultural research institutions. The transformation of a common resource into a commodity, of a self-regenerative resource into mere 'input' changes the nature of the seed and of agriculture itself. Since it robs peasants of their means of livelihood, the new technology becomes an instrument of poverty and underdeveloped.

The decoupling of seed from grain also changes the status of seed. From being a finished produce which rises from itself, nature's seeds and people's seeds become mere raw material for the production of corporate seed as commodity. The cycle of regeneration of biodiversity is therefore replaced by a linear flow of free germ plasm from farms and forests into labs and research stations, and the flow of modified uniform products as priced commodities from corporations to farmers. Diversity is destroyed by transforming it into mere raw material for industrial production based on uniformity which necessarily displaces the diversity of local agricultural practise.

Cropping systems in general, involve an interaction between soil, water and plant genetic resources. In indigenous agriculture, for example, cropping systems include a symbiotic relationship between soil, water, farm animals and plants. Green Revolution agriculture replaces this integration at the level of the farm with the integration of inputs such as seeds and chemicals. The indigenous cropping systems are based only on internal organic inputs. Seeds come from the farm, soil fertility comes from the farm and pest control is built into the crop mixtures. In the Green Revolution package, yields are intimately led to purchased inputs of seeds chemical fertilizers, pesticides and petroleum and intensive irrigation. High yields are not intrinsic to the seeds, but are a function of the availability of inputs. As referred to in Chapter 2 the UNRISD (United Nations Research Institute for Social Developments) 15 nation study of the impact of the new seeds concluded the term 'High Yielding Variety' is a misnomer because it implies that the new seeds are high-yielding in and of themselves. The distinguishing feature of the seeds, however, is that they are highly responsive to certain key inputs such as fertilizer and irrigation. Palmer therefore suggested the term 'high-responsive varieties' (HRVs) in place of high yielding varieties (HYV).[8]

As Claude Alvares has said 'for the first time the human race has produced seed that cannot cope on its own, but needs to be placed within an artificial environment for growth and output'.[9]

This change in the nature of seed is justified by creating a framework that treats self-regenerative seed as 'primitive' and as 'raw' germ plasm, and the seed that is inert without inputs and non-reproducible, as a finished product. The whole is rendered partial, the partial is rendered whole. The commoditised seed is ecologically incomplete and ruptured at two levels.

(1) It does not **reproduce** itself, while by definition, seed is a regenerative resource. Genetic resources are thus, through technological transformation, transformed from a renewable into a non renewable.

(2) It does not **produce** by itself. It needs the help of inputs to produce. As the seed and chemical companies merge, the dependence on inputs will increase, not decrease. And ecologically whether a chemical is added externally or internally, it remains an external input in the ecological cycle of the reproduction of seed.

It is this shift from the ecological processes of reproduction to the technological processes of production that underlies both the problem of dispossession of farmers and of genetic erosion.

The new plant biotechnologies will follow the path of the earlier HYV's of the Green Revolution in pushing farmers onto a technological treadmill. Biotechnology can be expected to increase the release of farmers on purchased inputs even as it accelerates the process of polarisation. It will even increase the use of chemicals instead of decreasing it. The dominant focus of research in genetic engineering is not on fertilizer-free and pest-free crops, but pesticide and herbicide-resistant varieties. For the seed-chemical multinations, this might make commercial sense, since it is cheaper to adapt the plant to the chemical than to adopt the chemical to the plant. The cost of developing a new crop variety rarely reaches $2 million, whereas the cost of a new herbicide exceeds $40 million.

As discussed in Chapter 6 herbicide and pesticide resistance will also increase the integration of seeds/chemicals and the multinationals' control of agriculture. A number of major agricultural-chemical companies are developing plants with resistance to their brand of herbicides. Soya beans have been made resistant to Ciba-Geigy's Atrazine herbicides, and this has increased annual sales of the herbicide by $120 million. Research is also being done to develop crop plants resistant to other herbicides, such as Dupont's 'Gist' and 'Glean' and Monsanto's 'Round-Up', which are lethal to most herbaceous plants and thus cannot be applied directly to crops. The successful development and sale of crop plants resistant to brand-name herbicides will result in further economic concentration of the agriculture industry market, increasing the market power of transnational companies. The farmers will own the land, but the corporation will own the crop in the field, giving instructions by a computer that monitors the progress and needs of a crop grown from genetically programmed seed.

Biotechnology can thus become an instrument for dispossessing the farmer of seed, as a means of production. The relocation of seed production from the farm to the corporate lab relocates power and value between the North and South; and between corporations and farmers. It is estimated that the elimination of home grown seed would dramatically increase the farmer's dependence on biotech industries by about $6 billion annually.

It also becomes an instrument of dispossession by selectively removing those plants or parts of plants that do not serve the commercial interest but are essential for survival of nature and people. 'Improvement' of a selected characteristic in a plant, is also a selection against other characteristics which are useful to nature, or for local consumption. Improvement is not a class or gender neutral concept. Improvement of partitioning efficiency is based on the 'enhancement of the yield of desired product at the expense of unwanted plant parts.' The desired product is however, not the same for rich people and poor people, or rich countries and poor countries, nor is efficiency. On the input side, richer people and richer countries are short of labour and poorer people and poorer countries are short of capital and land. Most agricultural development, however, increases capital input while displacing labour, thus destroying livelihoods. On the output side, which parts of a farming system of a plant will be treated as 'unwanted' for the better off may be the wanted part for the poor. The plants or 'plant parts' which serve the poor are the ones whose supply is squeezed by the normal priorities of improvement in response to commercial forces.

In the Indian context, plants that have been displaced by plant improvement in the Green Revolution are pulses and oilseeds, which are crucial to the nutrient needs of people and the soil. Monocultures of wheat and rice spread through the Green Revolution have also turned useful plants into weeds, as is the case with green leafy vegetables which grow as associates. Herbicides use has destroyed plants useful for the poor, pesticide use has destroyed the fish culture usually associated with paddy cultivation in Asian rice farming system. These losses through biodiversity destruction resulting from increasing yields of monocultures is never internalised in the productivity measure of technological change. Instead, both increased inputs and decreased outputs are externalised in measurement of productivity. Productivity is a measure of output per unit input. A typical subsistence farm in an Asian village produces more than twenty crops and supports animals. Individual crops are also multipurpose. Rice, for instance, is only partially food. The residue after the removal of the grain is not a nuisance to be disposed of as it would be on a farm in the developed world. The straw is used to feed work animals, to cook food, or even to help fertilize the field for the next year's crops – the straw and husk are important construction material. Traditional rice varieties produced five times as much straw as grain, and were an important source of food, feed, fuel, fertilizer and housing material.

Plant breeders, however, saw rice only as food and created a science and technology to increase grain yields. Traditional crop varieties, characterised by tall and thin straw typically convert the heavy doses of fertilizers into overall growth of the plant, rather than increasing the grain yield. Commonly, the excessive growth of the plant cause the stalk to break, lodging the grain on the ground, which results in heavy crop losses. The 'miracle seeds' or 'high yielding varieties' which started the process of the Green Revolution were biologically engineered to be dwarf varieties. The important feature of these new varieties is not that they are particularly productive in themselves but that they can absorb three or four times higher doses of fertilizer than the traditional varieties and convert it into grain, provided proportionately heavy and frequent irrigation applications are also available. They also have high susceptibility to insect and pest attacks.

For equivalent fertilization, the high yielding varieties produce about the same total biomass as the traditional rice. They increase the grain yield at the straw. Thus while traditional rice produces four to five times as much straw as grain, high yielding rice typically produce a one-to-one ratio of grain to straw. Thus a conversion from traditional to high yielding rice increases the grain available but decreases the straw. Abundance creates scarcity. Output as total biomass does not increase, but the input increases dramatically. High yielding varieties of wheat for example need about three times as much irrigation as traditional varieties. If water is considered as a critical output, the productivity of the new seeds is only a third of the productivity of traditional varieties. In terms of water use, the green revolution is clearly counter-productive. Increased irrigation intensity has future costs associated with it in of water logging and salinity, as the experience in India's Punjab has shown. In other regions such as Maharashtra and Tamil Nadu, the Green Revolution is causing large scale of ground water. Large scale dessication of the regions of the world now supporting the green revolution is thus a real possibility. Abundance again generates scarcity.

The increased water, fertilizer, and pesticide use is also counter-productive for the Asian farmer in financial terms. From studies conducted at the International Rice Research Institute, it is estimated that whereas the total cash costs of production for the average Filipino rice farmer using traditional methods and varieties is about $20 per hectare, the cost rises to $220 when the new, high yielding varieties are grown.

The fragmentation of components of the farm ecosystem and their integration with distant markets and industries is a characteristic of modern 'scientific' agriculture. The most common justification of the introduction of this system of food production is that it raises agricultural productivity. The high 'productivity' of modern agriculture is however a myth when total resource inputs are taken into account. The social and ecological costs with respect to the manufacture and use of fertilizers, pesticides and labour replacing energy and equipment are never taken into account, thus rendering the system artificially productive. If the energy used to provide all inputs to modern farming are deducted from the food calories produced modern agricultural technologies are found to be counterproductive. Whereas at the turn of the century even in the countries of the North, one calorie of food value was produced by the input of less than a calorie of energy so that there was a net gain, today ten calories of energy are used to produce the same one calorie of food value.

The higher productivity and efficiency of industrial agricultural is contextually determined, by selecting those inputs and outputs which suit the endowments of rich people and rich countries. Before we push through new agricultural technologies, it would be wise to pause and look at the other paths that were available for increased production but were never considered. Alternatives in agriculture have been based on conserving nature and peoples livelihoods, while improving yields, not on destroying them.

The myth that only chemically intensive labour displacing agriculture is productive has recently been challenged by a major study of the National Research Council in the U.S.[10] The report, titled 'Alternative Agriculture' shows that besides reducing the health and environmental hazards posed by chemical agriculture, alternative farming systems have economic viability. Traditionally, most evaluations comparing chemical farming to alternative practises have focussed principally on the cost and returns of adopting a specific farming method. Fewer studies have considered the impact of alternative farming systems on the economic performance of the whole farm. The committee could find no useful studies of the potential effects of widespread adoption of alternative agricultural systems. And the committee could find no aggregate studies that compare the costs and benefits of conventional agriculture with successful alternative systems. Most studies have the flawed approach of comparing the conventional farming practises with the economic performance of a similar farm, assuming withdrawal of certain categories of farm inputs instead of comparison with a farm with alternatives. The bias in favour of Green Revolution and chemical agriculture has distorted the assessment and potential of alternative agriculture. However, the new search for alternatives is showing that sustainable farming systems do not have to compromise at the level of productivity and yields.

A case study of the Thompson Farm established yields of 130-150 bushels/acre for corn against the national average of 124 bushels/acre, 45–55 bushels/acre of soyabean against the national average of 40 bushels/acre and 4 – 5 tonnes/acre of hay against 3–4 tonnes/acre of the national average. Similarly at the Kitamira farm, the yields of 35.5 tonnes/acre of tomatoes was much higher than the country's average.

It is not only countries like US which are making a shift away from energy and chemical intensive agriculture.

In India, a decade long experiment with 'rishi kheti' or alternative agriculture undertaken at Friends Rural Centre in Rasulia, Madhya Pradesh by Pratap Aggarwal yielded higher returns with indigenous seeds and no external inputs than the Green Revolution agriculture of HYV seeds and heavy chemical and irrigation inputs which had been earlier practised on the farm.[11]

In the terai-region of Uttar Pradesh a local farmer called Inder Singh continued to grow traditional varieties while most farmers shifted to a HYV called Pant-4. As intensive irrigation led to a decline in the water table, the thirsty HYVs could no longer be cultivated. Inder Singh's seeds came in useful. Owing to high productivity and low costs of cultivation in terms of fertilizer and water input, a particular variety called Indarasan (named after Inder Singh) spread to cover nearly 50% of the area. During drought years, Indarasan has coped much better than Pant-4, with yields of 32 quintals per acre compared to Pant-4 which has stagnated at 20 quintals per acre under irrigation. Without irrigation the HYV is totally destroyed. The Indarasan variety is also more remunerative in the market, selling at Rs208 per quintal compared to Rs175 per quintal for Pant-4.[12]

In Philippines on 29 May, 1986, the MASIPAC Centre (Farmers and Scientists for Agricultural Science Development Centre) was inaugurated in Jaen, Nueva Ecya. The centre's programme is to build an agricultural alternative which is not dependent on capital and external inputs as the Green Revolution was.[13]

Worldwide examples of successful alternative agriculture exist and are growing, even while they continue to be ignored by the dominant world view of agriculture. And it is these initiatives that carry the seeds of a sustainable agriculture. Blindness to these alternatives is not a proof of their non-existence. It is merely a reflection of the blindness.

Biotechnology Development and Biodiversity Conservation

The central paradox posed by the Green Revolution and biotechnology development is that modern plant improvement has been based on the destruction of the biodiversity which it uses as raw material. The irony of plant and animal breeding is that it destroys the very building blocks on which the technology depends. When agricultural modernisation schemes introduce new and uniform crops into the farmers' fields, they push into extinction the diversity of local varieties. In the words of Professor Garrison Wilkes of the University of Massachusetts, it is analogous to taking stones from a building's foundation to repair the roof.

And as Brian Ford-Lloyd and Michael Jackson elaborate,

'Current international activities surrounding the genetic resources of plants aim to confront one paradoxical prob-

lem. This is that scientists throughout the world are rightly engaged in developing better and higher yielding cultivars of crop plants to be used on increasingly larger scales. But this involves the replacement of the generally variable, lower yielding, locally adopted strains grown tradition-ally, by the products of modern agriculture – the case of uniformity replacing diversity. It is here that we find the paradox, for these self-same plant breeders are dependent upon the availability of a pool of diverse genetic material for success in their work. They are themselves dependent upon that which they are unwittingly destroying.'[14]

The paradox arises from the foundational errors of as-signment of value and utility, which then make the 'modern' varieties look inherently superior, whereas they are superior only in the context of increased control over plant genetic resources and a restricted production of certain commodities for the market.

The challenge of the 1990s is based on our getting rid of false notions of obsolescence and productivity which legit-imise the extinction of large parts of nature and society. The push for homogenisation and uniformity comes both from the transnational corporate sector, which has to create uni-formity to control markets. It also comes from the nature of modern research systems which have grown in response to the market. Since most biotechnology research is dictated by TNCs, the sought out solutions must have a global and homogenous character. TNCs do not tend to work for small market niches, but aim at large market shares. In addition, researchers prefer tasks that can be simplified enough to be tackled systematically, and that produce stable and widely applicable outcomes. Diversity goes against the standardisa-tion of scientific research.

However, more production of partial outputs as measured on the market and in a monoculture is often less production when measured in the diversity of nature's economy or people's sustenance economy. In the context of diversity, increased production and improved productivity can be consonant with biodiversity conservation. In fact it is often dependent on it.

There is, however, a prevalent misconception that biotechnology development will automatically lead to biodiversity conservation. The main problem with viewing biotech as a miracle solution to the biodiversity crisis is related to the fact that biotechnologies, are, in essence, technologies for the breeding of uniformity in plants and animals. Biotech corporations do talk of contribution to genetic diversity.

As John Duesing of Cuba Geigy states,

'Patent Protection will serve to stimulate the development of competing and diverse genetic solutions with access to these diverse solutions ensured by free market forces at work in biotech ecology and seed industries.'[15]

However, the 'diversity' of corporate strategies and the diversity of life-forms on this planet are not the same thing, and corporate competition can hardly be treated as a substitute for nature's evolution in the creation of genetic diversity.

Corporate strategies and products can lead to diversification of commodities, they cannot enrich nature's diversity. This confusion between commodity diversification and biodiversity conservation finds its parallel in raw-material diversification. Although breeders draw genetic materials from many places as raw material input, the seed commodity that is sold back to farmers is characterised by uniformity. Uniformity and monopolistic seed supplies go hand in hand. When this monopolising control is achieved through the molecular mind, destruction of diversity becomes more accelerated.

As Jack Kloppenburg has warned,

'Though the capacity to move genetic material between species is a means for introducing additional variations, it is also a means for engineering genetic uniformity across species.'[16]

Production is thus driven into the direction of destruction of diversity, which swallows all biodiversity into its domain of colonisation. Production based on uniformity becomes the primary threat to biodiversity conservation, even though in the convoluted political economy of the market, it is cited as the reason for conservation. 'The exploitation of genetic diversity for crop improvement should be the ultimate objective of genetic resources exploration and conservation,' it is argued.

The arbitrary inequality created in the status of germ plasm, creates an arbitrary separation between production and conservation. Some people's germplasm becomes a finished commodity, a 'product', other people's germplasm becomes mere 'raw' material for that product. The manufacture of the 'product' in corporate labs is counted as production. The reproduction of the 'raw' material by nature and Third World farmers is mere conservation. The 'value added' in one domain is built on the 'value robbed' in other domain. Biotechnology development thus translates into biodiversity erosion and poverty creation.

The main challenge to biodiversity conservation is the removal of reductionist blinkers which make more look less and less look more. The social construction of this deceptive 'growth and productivity' is achieved by:

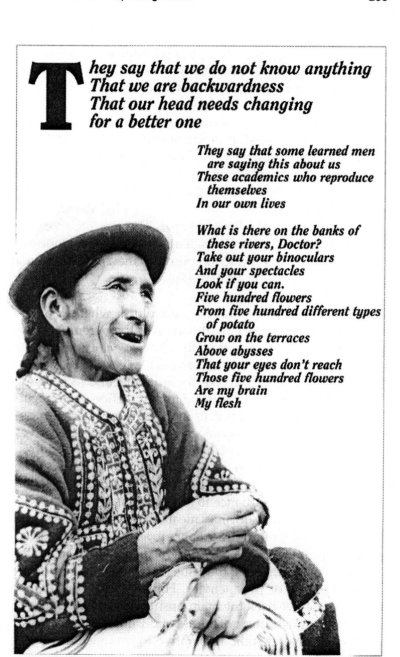

They say that we do not know anything
That we are backwardness
That our head needs changing
for a better one

They say that some learned men
* are saying this about us*
These academics who reproduce
* themselves*
In our own lives

What is there on the banks of
* these rivers, Doctor?*
Take out your binoculars
And your spectacles
Look if you can.
Five hundred flowers
From five hundred different types
* of potato*
Grow on the terraces
Above abysses
That your eyes don't reach
Those five hundred flowers
Are my brain
My flesh

From **A call to certain academics** by Jose Maria Arguedas. Translated from the Quechua by William Rowe.

(1) excluding crops and parts of crops as 'unwanted'
(2) creating a false hierarchy of resources and knowledge, splitting diversity into dichotomy.

Not till diversity is made the logic of production can diversity be conserved. If production continues to be based on the logic of uniformity and homogenisation, uniformity will continue to displace diversity. 'Improvement' from the corporate viewpoint, or from the viewpoint of western agricultural research is often a loss for the Third World, especially the poor in the Third World. There is therefore no inevitability that production acts against diversity. Uniformity as a pattern of production becomes inevitable only in a context of control and profitability.

All systems of sustainable agriculture whether of the past or the future work on the basis of the perennial principles of diversity and reciprocity. The two principles are not independent but interrelated. Diversity gives rise to the ecological space for give and take, for mutuality and reciprocity. Destruction of diversity is linked to the creation of monocultures, and with the creation of monocultures, the self-regulation and decentred organisation of diverse systems gives way to external inputs and external and centralised control. Sustainability and diversity are ecologically linked because diversity offers the multiplicity of interactions which can heal ecological disturbance to any part of the system. Nonsustainability and uniformity means that disturbance to one part is translated into a disturbance to all other parts. Instead of being contained, ecological destabilisation tends to be amplified. Closely linked to the issue of diversity and uniformity is the issue of productivity. Higher yields and higher production have been the main push for the introduction of uniformity and the logic of the assembly line. The imperative of growth generates the imperative for monocultures. Yet this growth is, in large measure, a socially constructed, value laden category. It exists as a 'fact' by excluding and erasing the facts of diversity and production through diversity. Sustainability, diversity and decentred self-organisation are therefore linked, as are unsustainability, uniformity and centralisation.

Diversity as a pattern of production, not merely of conservation, ensures pluralism and decentralisation. It prevents the dichotomising of biological systems into 'primitive' and 'advanced'. Like Gandhi challenged the false concepts of obsolescence and productivity in the production of textiles by his search for the spinning wheel, groups across the Third World are challenging the false concepts of obsolescence in agricultural production by searching for seeds used by farmers over centuries and making them the basis of a futuristic self-reliant and sustainable agriculture.

Patents, Intellectual Property and the Politics of Knowledge

Like the spinning wheel was rendered backward and obsolete by an earlier technological revolution, farmers' seeds are rendered incomplete and valueless by the process that makes corporate seeds the basis of wealth creation. The indigenous varieties or land races, evolved through both natural and human selection, and produced and used by Third World farmers worldwide are called 'primitive cultivars'. Those varieties created by modern plant breeders in international research centres or by transnational seed corporations are called 'advanced' or 'elite'. The tacit heirarchy in words like 'primitive' and 'elite' becomes an explicit one in the process of conflict. Thus, the North has always used Third World germplasm as a freely available resource and treated it as valueless. The advanced capitalist nations wish to retain free access to the developing world's storehouse of genetic diversity, while the South would like to have the proprietory varieties of the North's industry declared a similarly 'public' good. The North, however, resists this democracy based on the logic of the market. Williams, Executive Secretary of IBPGR has argued that, 'It is not the original material which produces cash returns'.

A 1983 forum on Plant Breeding, sponsored by Pioneer Hi-Bred stated that,

> 'Some insist that since germplasm is a resource belonging to the public, such improved varieties would be supplied to farmers in the source country at either or no cost. This overlooks the fact that 'raw' germplasm only becomes valuable after considerable investment of time and money, both in adapting exotic germplasm for use by applied plant breeders in incorporating the germplasm into varieties useful to farmers.'[17]

The corporate perspective views as value only that which serves the market. However, all material processes also serve ecological needs and social needs, and these needs are undermined by the monopolising tendency of corporations.

The issue of patent protection for modified life forms raises a number of unresolved political questions about ownership and control of genetic resources. The problem is that in manipulating life forms you do not start from nothing, but from other life-forms which belong to others – maybe through customary law. Secondly, genetic engineering and biotechnology does not create new genes, it merely relocates genes already existing in organisms. In making genes the object of value through the patent system, a dangerous shift takes place in the approach to genetic resources.

Most Third World countries view genetic resources a

common heritage. In most countries animals and plants were excluded from the patent system until recently when the advent of biotechnologies changed concepts of ownership of life. With the new biotechnologies life can now be owned. The potential for gene manipulation reduces the organism to its genetic constituents. Centuries of innovation are totally disvalued to give monopoly rights on life forms to those who manipulate genes with new technologies, placing their contribution over and above the intellectual contribution of generations of Third World farmers for over ten thousand years in the areas of contribution, breeding, domestication and development of plant and animal genetic resources.

As Pat Mooney has argued,

'the argument that intellectual property is only recognisable when performed in laboratories with while lab coats is fundamentally a racist view of scientific development.'[18]

Two biases are inherent in this argument. One, that the labour of Third World farmers has no value, while labour of Western scientists adds value. Secondly, that value is a measure only in the market. However, it is recognised that, 'the total genetic changed achieved by farmers over the millenia was far greater than that achieved by the last hundred or two years of more systematic science-based efforts'.[19] Plant scientists are not the sole producers of utility in seed.

This utility of farmers seeds has high social and ecological value, even if it has no market value attached.to it. The limits of the market system in assigning value can hardly be a reason for denying value to farmers' seeds. It points more to the deficiency of the logic of the market than the status of the seed or the farmers intellect.

There is no epistemological justification for treating some

germplasm as valueless and common and other germplasm as a valuable commodity and private property. This distinction is not based on the nature of the germ plasm, but on the nature of political and economic power.

Putting value on the gene through patents makes biology stand on its head. Complex organisms which have evolved over millenia in nature, and through the contributions of Third World peasants, tribals and healers are reduced to their parts, and treated as mere inputs into genetic engineering. Patenting of gene thus leads to a devaluation of life-forms by reducing them to their constituents and allowing them to be repeatedly owned as private property. This reductions and fragmentation might be convenient for commercial concerns but it violates the integrity of life as well as the common property rights of Third World peoples. On these false notions of genetic resources and their ownership through intellectual property rights are based the 'bio-battles' at FAO and the trade wars at GATT. Countries like US are using trade as a means of enforcing their patent laws and intellectual property rights on the sovereign nations of the Third World. US has accused countries of the Third World as engaging in 'unfair trading practice' if they fail to adopt US patent laws which allow monopoly rights in life form. Yet it is the US which has engaged in unfair practices related to the use of Third World genetic resources. It has freely taken the biological adversity of the Third World to spin millions of dollars of profits, none of which have been shared with Third World countries, the original owners of the germplasm. A wild tomato variety (Lycopresicon chomrelweskii) taken from Peru in 1962 has contributed $8 million a year to the American tomato processing industry by increasing the content of soluble solids. Yet none of these profits or benefits have been shared with Peru, the original source of the genetic material.

According to Prescott-allen, wild varieties contributed US$340 million per year between 1976 and 1980 to the US farm economy. The total contribution of wild germ plasm to the American economy has been US$66 billion, which is more than the total international debt of Mexico and Philippines combined. This wild material is 'owned' by sovereign states and by local people.[19]

Patents and intellectual property rights are at the centre of the protection of the right to profits. Humans rights are at the centre of the protection of the right to life, which is threatened by the new biotechnologies, which are expanding the domain of and drive for capital accumulation, while introducing new risks and hazards for citizens.

The words 'freedom' and 'protection' have been robbed of their humane meaning and are being absorbed into the double-speak of corporate jargon. With double-speak are associated double standards, one for citizens and one for corporations, one for corporate responsibility and one for corporate profits.

The USA is the most sophisticated in the practice of double standards and the destruction of people's rights to health and safety in the Third World. On the one hand it aims at keeping regulation for safeguards restricted to its own geographical boundaries, while on the other hand it aims at destroying the Indian Patents Act of 1970 and replacing it with a strong US-style system of patent protection which is heavily biased in favour of the industrially developed countries.

The US government considers the transnationals' lack of patent protection as unfair trading practice. It does not consider the destruction of regulation for public safety and environmental protection as unethical and unfair for the citizens of the Third World. The USA wants to limit and localise laws for the protection of people and universalise laws for the protection of profits. The people of India want the reverse – a universalisation of the safety regulations protecting people's right to life and livelihoods and a localisation of laws relating to intellectual property and private profits.

All life is precious. It is equally precious to the rich and the poor, the white and the black, to men and women. Universalization of the protection of life is an ethical imperative. On the other hand, private property and private profits are culturally and socio-economically legitimised constructs holding only for some groups. They do not hold for all societies and all cultures. Laws for the protection of private property rights, especially as related to life forms, cannot and should not be imposed globally. They need to be restrained.

Double standards also exist in the shift from private gain to social responsibility for environmental costs. When the patenting of life is at issue, arguments from 'novelty' are used. Novelty requires that the subject matter of a patent be new, that it be the result of an inventive step, and not something existing in nature. On the other hand, when it comes to legislative safeguards, the argument shifts to 'similarity', to establishing that biotechnology products and genetically engineered organisms differ little from parent organisms.

To have one law for environmental responsibility and another for proprietary rights and profits is an expression of double standards. Double standards are ethically unjustified and illegitimate, especially when they deal with life itself. However, double standards are consistent with and necessary for the defence of private property rights. It is these double standards which allow the life and livelihoods of the people and the planet to be sacrificed for the protection of profits.

The resistance to such anti-life technological shifts requires that we widen the circle of control and decision-making about technology, by treating technology in its social and ecological context. By keeping human rights at the centre of discourse and debate on new technologies, we might be able to restrain the ultimate privatization of life itself.

During the first industrial revolution and its associated colonisation, Gandhi had transformed the 'primitive' spinning wheel into a living symbol of the struggle for India's freedom and self-determination. The 'primitive' seeds of Third World peasants could well become the symbols of the struggle for freedom and the protection of life in the emerging context of recolonisation of the Third World and its living resources.

References 7

1.   Vandana Shiva and J Bandyopadhyay, 'Political Economy of Technological Polarisations', *Economic and Political Weekly*, Vol. XVII, No. 45, 6 November, 1982, p1827-32.

2.   S Arasarathnam, 'Weavers, Merchants and Company: The Handloom Industry in South Eastern India', *The Indian Economic and Social History Review*, Vol. 17, No. 3, p281.

3.   J G Borpujari, 'Indian Cotton and the Cotton Famine, 1860-65', *The Indian Economic and Social History Review*, Vol. 10, No. 1, p45.

4.   D R Gadgil, *Industrial Revolution of India in Recent Times 1860-1939*, Bombay: Oxford University Press, 1971, p329.

5.   Quoted in Pyarelal, *Towards New Horizons*, Ahmedabad: Navjivan Press, 1959, p150.

6.   *Ibid.*

7.   Jack Kloppenburg, *First the Seed*, Cambridge (USA): Cambridge University Press, USA, 1988.

8.   Lappe & Collins, *op cit*, p114.

9.   Claude Alvares, 'The Great Gene Robbery', *The Illustrated Weekly of India*, 23 March, 1986.

10. National Research Council, Alternative Agriculture National Academy Press, Washington, DC, 1989.

11. Pratap C Aggarwal, ' Natural Farming Succeeds in Indian Village', *Return to the Good Earth*, Penang: Third World Network, 1990, p461.

12. Vir Singh and Satya Prakash, 'India farmers rediscover advantages of traditional rice varieties', *Return to the Good Earth*, Penang: Third World Network, 1990, p546.

13. Rolanda B Modina and A R Ridao, *IRRI Rice : The Miracle that never was*, ACE Foundation, Quezon City, Philippines, undated.

14. Brian Ford-Lloyd and Michael Jackson, *Plant Genetic Resources*, Edward Arnold, 1986, p1.

15. Statement of John Duesing, in meeting on Patents ail European Parliament, Brussels, Feb 1990.

16. Jack Kloppenburg, *op cit*.

17. Quote in Jack Kloppenburg, p185.

18. Pat Mooney, 'From Cabbages to Kings', *Intellectual Property vs Intellectual Integrity*, ICDA report.

19. Norman Summonds, *Principles of Crop Improvement*, New York: Longman, 1979, p11.

20. Hugh Iltis, 'Serendipity in Exploration of Biodiversity: What good are weedy tomatoes', in E O Wilson, (ed), *Biodiversity*, National Academy Press, 1986.